Lecture Notes in Computer Scie

T0230174

Commenced Publication in 1973
Founding and Former Series Editors:
Gerhard Goos, Juris Hartmanis, and Jan van Leeuwen

Sotiris Nikoletseas   José D.P. Rolim (Eds.)

Algorithmic Aspects
of Wireless
Sensor Networks

First International Workshop, ALGOSENSORS 2004
Turku, Finland, July 16, 2004
Proceedings

 Springer

Volume Editors

Sotiris Nikoletseas
University of Patras and Computer Technology Institute (CIT)
61 Riga Fereou Street, 26221 Patras, Greece
E-mail: nikole@cti.gr

José D.P. Rolim
Université de Genève, Centre Universitaire d'Informatique
24, Rue Général Dufour, 1211 Genève 4, Suisse
E-mail: jose.rolim@cui.unige.ch

Library of Congress Control Number: 200410882

CR Subject Classification (1998): F.2, C.2, E.1, G.2

ISSN 0302-9743
ISBN 3-540-22476-9 Springer-Verlag Berlin Heidelberg New York

Springer-Verlag is a part of Springer Science+Business Media

springeronline.com

© Springer-Verlag Berlin Heidelberg 2004

Printed in Germany

Typesetting: Camera-ready by author, data conversion by PTP-Berlin, Protago-TeX-Production GmbH
Printed on acid-free paper   SPIN: 11019060   06/3142   5 4 3 2 1 0

Preface

This volume contains the contributed papers and invited talks presented at the 1st International Workshop on Algorithmic Aspects of Wireless Sensor Networks (ALGOSENSORS 2004), which was held July 16, 2004, in Turku, Finland, co-located with the 31st International Colloquium on Automata, Languages, and Programming (ICALP 2004).

Wireless ad hoc sensor networks have become a very important research subject due to their potential to provide diverse services in numerous applications. The realization of sensor networks requires intensive technical research and development efforts, especially in power-aware scalable wireless ad hoc communications protocols, due to their unusual application requirements and severe constraints.

On the other hand, a solid theoretical background seems necessary for sensor networks to achieve their full potential. It is an algorithmic challenge to achieve efficient and robust realizations of such large, highly dynamic, complex, non-conventional networking environments. Features, including the huge number of sensor devices involved, the severe power, computational and memory limitations, their dense deployment and frequent failures, pose new design, analysis and implementation challenges.

This event is intended to provide a forum for researchers and practitioners to present their contributions related to all aspects of wireless sensor networks.

Topics of interest for ALGOSENSORS 2004 were:

- Modeling of specific sensor networks.
- Methods for ad hoc deployment.
- Algorithms for sensor localization and tracking of mobile users.
- Dynamic sensor networks.
- Hierarchical clustering architectures.
- Attribute-based named networks.
- Routing: implosion issues and resource management.
- Communication protocols.
- Media access control in sensor networks.
- Simulators for sensor networks.
- Sensor architecture.
- Energy issues.

This volume contains 2 invited papers related to corresponding keynote talks, one by Viktor Prasanna (University of Southern California, USA) and one by Paul Spirakis (University of Patras and Computer Technology Institute, Greece) and 15 contributed papers that were selected by the Program Committee (PC) from 40 submitted papers. Each paper was reviewed by at least 2 PC members, while a total of 99 reviews were solicited.

We would like to thank all the authors who submitted papers to ALGOSENSORS 2004, the members of the Program Committee, as well as the external referees. Also we thank the members of the Organizing Committee. We especially wish to thank Prof. Dr. Jan van Leeuwen for valuable comments.

We gratefully acknowledge the support from the Research Academic Computer Technology Institute (RACTI, Greece, http://www.cti.gr), and the Athens Information Technology (AIT, Greece, http://www.ait.gr) Center of Excellence for Research and Graduate Education. Also, we thank the European Union (EU) IST/FET ("Future and Emerging Technologies") R&D Projects of the Global Computing (GC) Proactive Initiative FLAGS (IST-2001-33116, "Foundational Aspects of Global Computing Systems") and CRESCCO (IST-2001-33135, "Critical Resource Sharing for Cooperation in Complex Systems") for supporting ALGOSENSORS 2004. Finally, we wish to thank Springer-Verlag, Lecture Notes in Computer Science (LNCS), and in particular Alfred Hofmann, as well as Anna Kramer and Ingrid Beyer, for a very nice and efficient cooperation.

July 16, 2004   Sotiris Nikoletseas and Jose Rolim

Organization

Program and General Committee Co-chairs

Sotiris Nikoletseas   Patras University and Computer Technology Institute, Greece
Jose Rolim   Geneva University, Switzerland

Program Committee

Ian Akyildiz   Georgia Institute of Technology, USA
Azzedine Boukerche   University of Ottawa, Canada
Deborah Estrin   University of California Los Angeles, USA
Afonso Ferreira   CNRS, I3S Inria Sophia Antipolis, France
Alfredo Ferro   University of Catania, Italy
Wendi Heinzelman   University of Rochester, USA
Chalermek Intanagonwiwat   Chulalongkorn University, Thailand
Elias Koutsoupias   University of Athens, Greece
Bhaskar Krishnamachari   University of Southern California, USA
Stefano Leonardi   University of Rome "La Sapienza", Italy
Sotiris Nikoletseas   University of Patras and CTI, Greece
Viktor Prasanna   University of Southern California, USA
Jose Rolim   University of Geneva, Switzerland
Peter Sanders   Max Planck Institute for CS, Germany
Maria Serna   Polytechnic University of Catalunya, Spain
Christian Schindelhauer   University of Paderborn, Germany
Paul Spirakis   University of Patras and CTI, Greece
Peter Triantafilloy   University of Patras and CTI, Greece
Eli Upfal   Brown University, USA
Jennifer Welch   Texas A&M University, USA
Peter Widmayer   ETH Zurich, Switzerland

Organizing Committee

Ioannis Chatzigiannakis   Patras University and CTI, Greece, Chair
Charilaos Efthymiou   Patras University and CTI, Greece
Athanasios Kinalis   Patras University and CTI, Greece

Referees

Ioannis Aekaterinidis	Suhas Diggavi	Stefan Ruehrup
Amol Bakshi	Rosalba Giugno	Mirela Sechi Moretti
Luca Becchetti	Bo Hong	Annoni Notare
Domenico Cantone	Michelle Liu Jing	Stanislava Soro
Guangtong Cao	Nicholas Neumann	Bulent Tavli
Ioannis Chatzigiannakis	Nikos Ntarmos	Andrea Vitaletti
Lei Chen	Tolga Numanoglu	Klaus Volbert
Yu Chen	Paolo Penna	Owen Zacharias
Zhao Cheng	Mark Perillo	Yan Yu
Gianluca Cincotti	Alfredo Pulvirenti	Yang Yu

Sponsoring Institutions

- Research Academic Computer Technology Institute (RACTI), Greece
- Athens Information Technology (AIT), Athens, Greece
- EU-FET R&D Project "Foundational Aspects of Global Computing Systems", (FLAGS)
- EU-FET R&D Project "Critical Resource Sharing for Cooperation in Complex Systems", (CRESCCO)

Table of Contents

Algorithm Design and Optimization for Sensor Systems*
(Invited Talk)

Viktor K. Prasanna

Department of Electrical Engineering
3740 McClintock Ave., EEB-200C
University of Southern California
Los Angeles CA 90089-2562 USA
prasanna@usc.edu
http://ceng.usc.edu/~prasanna

Distributed collaborative computation will play a key role in next generation of smart sensor systems. Till now, most research has focused upon development of networking protocols and simple data aggregation algorithms to facilitate robust communication and localized processing in sensor systems. One of the key challenges involved in realizing future sensor systems is design of efficient and scalable algorithms to facilitate large scale collaborative computation in such networks.

In the state-of-the-art, partitioning of computation tasks among nodes, node-level optimization, and data routing is done in an ad-hoc, largely empirical manner. While this does not necessarily result in inferior designs, it does require the application developer to be aware of the details of the underlying node hardware, networking paradigms, and their performance. We believe that next logical step in the evolution of networked wireless sensor systems is to develop computation models and programming abstractions to complement the existing body research. A layer of abstraction from an algorithm design perspective is needed to provide the end user with a modular, layered, composable paradigm to design and optimize networked sensor systems.

In this talk, we discuss various issues in addressing the following questions:

- What are the suitable abstractions for networked sensor systems that can be used by a programmer to develop applications that are efficient with respect to relevant performance metrics?
- Given control over the lower layers of the protocol stack, and also over hardware 'knobs' such as radio range, power states, etc., how to design and optimize algorithms so that optimal or near-optimal time and energy performance can be realized?
- What are the reusable primitives for rapid application synthesis?

---

* This work is supported in part by the US National Science Foundation under award No. IIS-0330445.

S. Nikoletseas and J. Rolim (Eds.): ALGOSENSORS 2004, LNCS 3121, pp. 1–2, 2004.

Although we approach sensor systems from a parallel and distributed systems' perspective, most of the general models of parallel and distributed computation need to be redefined. Wireless communication and energy constraints are two major factors that are responsible for this.

To illustrate our ideas, we consider a network of sensor nodes uniformly distributed over a terrain, continuously sampling the environment. We define a simple model of computation to evaluate the performance of various techniques for the classic problem of topographic querying. Topographic querying is the process of extracting metadata from a sensor network to understand the graphic delineation of regions of interest. We propose a simple technique that creates a hierarchical distributed storage infrastructure in the sensor system and performs in-network aggregation at each level of the hierarchy to reduce the total energy and latency. We conclude by summarizing issues in automatic synthesis of sensor systems.

Algorithmic and Foundational Aspects of Sensor Systems*
(Invited Talk)

Paul G. Spirakis

Computer Technology Institute (CTI) and University of Patras,
P. O. Box 1122, 261 10 Patras, Greece
spirakis@cti.gr

Abstract. We discuss here some algorithmic and complexity-theoretic problems motivated by the new technology advances in sensor networks and smart dust. We feel that the field of algorithms has a nice and sizeable intersection with the field of design and control of sensor networks. New measures of efficiency and quality of such algorithms are discussed. Some clear directions for future research are highlighted.

1 Introduction

Sensor networks employ a set of a vast number of very small, fully autonomous computing and communication devices. These devices (the network's *nodes*) should co-operate in order to accomplish a *global* task.

When the devices are ultra-small (abstracted to *points*) then the name *smart dust* (or *smart dust cloud*) is used. We use the term *particles* then for nodes.

The set of nodes of such a network is usually assumed to take *ad hoc* positions in e.g. a surface (2 dimensions) or in an area in the 3-dimensional space. The *area* (or space) *of deployment* of the sensors may include *obstacles* (or *lakes*) i.e. subareas where no sensing device can be found.

Sensor nets differ from general ad-hoc nets with wireless communication in the sense that local resources of each node (such as available energy, storage, communication/computation capability, reliability) are seriously constrained. (See [1], [2] for a nice survey.) In most sensor networks there are also one (or more) *base stations* to which data collected by sensor nodes is relayed for uploading to external systems.

Although the technology of construction of individual cheap sensors is quite advanced today, (e.g. there are sensors of less than a cubic *cm* size, that can even sense a perfume's smell [3]), the corresponding system aspects research is still at its infancy. The general question of "what can a sensors net do/not do globally" is quite open and it is a challenge for the algorithmic thought. Abstract models

---

* This work has been partially supported by the IST Programme of the European Union under contract number IST-2001-33116 (**FLAGS**) and within the 6th Framework Programme under contract 001907 (DELIS).

S. Nikoletseas and J. Rolim (Eds.): ALGOSENSORS 2004, LNCS 3121, pp. 3–8, 2004.
© Springer-Verlag Berlin Heidelberg 2004

of such nets exist but there are many of them, each emphasizing certain aspects and connecting to certain fields of e.g. Mathematics (or Physics). For example, there is a whole field of *Random Geometric Networks* (see e.g. [5]) aiming at studying the combinatorial properties of, usually, geometric graphs created by a randomly deployed set of points in an area of the plane and having edges among two points when they are able to communicate (i.e. they happen to be close in a certain sense).

Thus, it seems that a (small maybe but technologically crucial) algorithmic subfield is trying to demonstrate itself, motivated by such networks/systems. We highlight here some concrete algorithmic problems of this field; the further development of the algorithmic thought here is, for us, a basic prerequisite for most of the pragmatic issues (software, middleware, protocols e.t.c.) aiming at controlling or *successfully* exploiting such systems.

Furthermore, it is obvious from the start that these very restricted systems cannot do everything; the ideas of impossibility results and/or trade-offs (e.g. arising from the foundations of distributed computing) should be welcomed here.

2   Information Propagation and the "Ad-Hoc" Notion

Perhaps the most natural problem in sensor systems is the "efficient" propagation of a sensed *local* event E towards some receiving center (assuming an event-driven data delivery operation). For example, consider the case where a node (sensor) in the field senses a local event (e.g. a local fire in the forest); such an event may arise arbitrarily and at any time, triggered by unusual changes in the local environment. The goal here is to inform the base station quickly, without of course depleting the net from its resources.

The difficulty here stems from (a) the ad-hoc position of nodes in the area "covered" by the sensors net (b) the fact that usually each sensor has its own coordinates system (c) the basic restriction that a sensor node v can communicate info(E) (the necessary information about the event E) only to nodes that happen to be "nearby" (within v's communication *radius* r). Even when each node r knows *a priori* the direction to the base center (*sink*) and even if nodes can broadcast within an angle of radius r, still info(E) will travel via a sequence of *hops* to the sink. In the (unrealistic) case in which a powerful Mind knew all the topology of the system, a *shortest path*, P, to the sink could be established. (Here "shortest" may include energy availability.)

Thus, the *hops efficiency* $h(A)$ of a local propagation protocol A can be defined to be the ratio of the number $l(A)$ of hops done by A to reach the sink, divided by the "length" (minimum number of hops) of P, $l(P)$. This is a *competitive* measure, motivated by the seemingly on-line nature of which node is nearby to which. The question is interesting even if one assumes a known probability distribution of sensors in the area; the existence of obstacles (or lakes) makes it even harder and resembles older research on searching "without a map" (see e.g. [6], [7]).

Assuming that each local execution of the *(forwarding) Propagation Protocol* A spends some considerable energy of the sensor, we conclude that a successful Propagation Protocol should not involve sensors that are "far away" from the shortest path from the source of the local event to the sink. (An obvious, flooding the net, protocol, will succeed in sending info(E) wherever possible but it will employ the whole net.) Thus, each node receiving info(E) has a basic, local, decision to make: to further propagate info(E) or not?

If n_A are the nodes that forward info(E) via protocol A and n are all the nodes in the sensors *field*, then $\frac{n_A}{n}$ may serve as another performance measure of A that abstracts the notion of energy efficiency of A.

Robustness of a propagation protocol to single sensors failures, as well as *scalability* (to any size n of the net) are two other important properties here. For some good proposals to design efficient propagation protocols, see e.g. [6], [4], [8].

The whole issue of efficient propagation of a local event to a source is connected to the issue of *Geometric Routing* [9]. Since, optimally, each node should deliver the packet to the closest (to the sink) neighbor, initially one may think of a simple *greedy* algorithm based on their principle. However, *voids* (nodes with no neighbor closer to the sink) may exist, due to the distribution of nodes or due to obstacles, lakes or failures.

Since GPS antennae are costly (both in price and energy consumption) it is unlikely that future ad-hoc nets can rely on the availability of precise geographic coordinates. [10] recently proposed the idea in which nodes first decide on *fictitious virtual coordinates* and then apply greedy routing based on those. But, even if the construction of such virtual coordinates can be allowed as a preprocessing step in the network, can such coordinates always exist? Which must be the properties of the network's graph in order to guarantee such coordinates?

In fact, if G is a graph with nodes embedded in $\mathcal{R}^k$ and d is the k-dimensional Euclidian distance, define a path of nodes $(v_0, v_1, \ldots, v_m)$ to be *distance decreasing* if $d(v_i, v_m) < d(v_{i-1}, v_m)$ for $i = 1, \ldots, m$.

If G has the property that $\forall s, t$ nodes of G *there exists* a distance decreasing path from s to t, and if also there exists an efficient way for the net G to construct these paths, then the greedy routing problem will be solved nicely. Relevant questions (and answers) here can be found in e.g. [9] but many versions of the problem are open.

3   Some Energy Optimality Considerations

The attenuation of a signal with power P_s at the source s (i.e. its power P_r at a node r) is

$$P_r = \frac{P_s}{d(s,r)^\delta}$$

where $d(s, r)$ is the Euclidian distance of s, r and $\delta \geq 2$ is the so called *distance-power gradient* [11].

Assume now that nodes in the sensor field can decide (based on an algorithm) on *how far* to broadcast (i.e. they can modify the radius of broadcasting). What is then the best distances to broadcast?

The problem is interesting even in the simple case of the net being an 1-dimensional lattice from source s to sink t (Fig. 1).

Fig. 1. 1-dimensional lattice from source s to sink t

Let l be the number of edges in the lattice (say, each of distance r). Consider the following decision problem for each node v_i: Should v_i broadcast just by radius r, or use a very big radius (e.g. an $R \geq d(v_i, t)$ would do)?

This problem always reminds the corresponding decision problem of a soccer player: to shoot directly, or to pass the ball to a next, closer to the target, attacker?

If a protocol A here makes initially x small hops and then a final broadcast to t, then the delivery time of A is $T = x + 1$, while the energy spent (assuming a unit of energy for each small hop) is $E = x + c(l - x)^\delta$ where c is a constant. But then

$$E = (T - 1) + c(l + 1 - T)^\delta \tag{1}$$

and this is a "universal" energy-time trade-off for all such protocols A. Such trade-offs (and their discovery) are reminiscent of the area-time trade-offs of the older research in VLSI, and their precise statement will be of immense value.

Of course, there are many refinements and complications to the above question. First of all, a message can be decoded only if P_r is no less than source threshold γ. But then, node v_i may not have enough energy left to make the final "long jump".

If we now think of a graph G instead of a line, and of the need of propagating a sequence of events to the sink t, then any successful protocol should face the problem of energy-depleted nodes in the time course of serving the whole sequence of events.

Furthermore, for any position of the sink t, its nearby (within communication distance) nodes are always bounded by some number related to the volume of the sphere of radius r around t. These nodes are bound to suffer severe energy losses even when the origins of local events are fairly distributed in the sensors field.

Such limitations lead to a sequence of problems (*Range assignment problems*) that can be viewed either as off-line optimization questions or even on-line questions. The complexity of such problems has been well-studied in the off-line setting (see e.g. [12]) but even there many questions remain.

4   Hybrid Networks

A smart dust cloud, covering (i.e. deployed) in some area, can be seen (at a high level) as just a uniform communication medium giving, between any two nodes in the area, a "direct" communication link of a certain capacity c.

Consider now the case of a given, fixed, network, abstracted as a communications graph $G(V, E)$. We might want to superimpose on a selected subset of nodes $V' \subseteq V$, a smart dust cloud covering exactly the nodes in V' (via, of course, a vast number of sensor nodes deployed in the area (subgraph) defined by V') (Fig. 2).

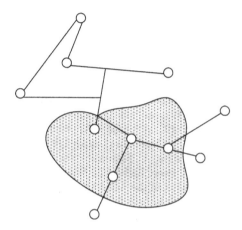

Fig. 2. Superposition of smart dust on fixed topology nets

In abstract terms this means that we overlap G with a clique in V' of edge capacity c. What are the new capabilities of the resulting net? For example, what is the maximum flow that can be pushed via G after the overlap? What is the best selection of V' (note that the induced graph $G(V')$ may be restricted to be connected or follow some other "proximity" restriction) to maximize the resulting flow? How about the connectivity (or the chromatic number) of the resulting graph?

Many of such questions can be easily seen to be NP-complete when we ask for optimal V' ([13]). The practicality of such problems comes from the easiness of deploying a "dense" random net of sensors in an area (e.g. just throw them from a plane) and such areas might be hard to intervene in order to carefully upgrade old existing networks (e.g. they are densely populated, hostile, or even simply dangerous).

The field of such overlaps of sensor clouds to existing nets is yet another promising research direction.

5 Conclusions

We presented here a set of algorithmic problems motivated by the arising technology of sensor networks. This set is partial, but if even so, we feel that a new algorithmic subfield demonstrates its existence. Its further development will certainly help in better ways of designing, controlling and using such networks. Limitations of sensor nets with respect to global computing tasks do exist and can be nicely demonstrated and characterized by the abstract thought in algorithms.

References

1. I. Akyildiz, S. Weilian, Y. Sankarasubramanian, E. Cayirici. "A Survey on Sensor Networks", IEEE Communications Magazine, August 2002.
2. Chee-Yee Chong, Kumar SP. "Sensor Networks: Evolution, Opportunities and Challenges", Proc. of the IEEE, August 2003.
3. Jorge Garcia Vidal. Personal Communication, 2004.
4. I. Chatzigiannakis, T. Dimitriou, S. Nikoletseas, P. Spirakis. "A Probabilistic Forwarding Protocol for Efficient Data Propagation in Sensor Networks", 5th European Wireless Conference (EW), pp. 344-350, Barcelona, 2004.
5. J. Diaz, M. D. Penrose, J. Petit, M. Serna. "Approximation layout problems on Random Geometric Graphs", J. Algorithms 39:78-116, 2001.
6. I. Chatzigiannakis, S. Nikoletseas, P. Spirakis. "Smart Dust Protocols for Local Detection and Propagation". Proc. 2nd ACM Workshop on Principles of Mobile Computing (POMC), 2002. Also, in the ACM Mobile Networks (MONET) Journal, Special Issue on Algorithmic Solutions for Wireless, Mobile, Ad hoc and Sensor Networks, accepted, to appear in 2004.
7. C. Papadimitriou, M. Yannakakis. "Shortest Paths without a Map", Theoretical Computer Science 84(1), pp. 127-150, 1991.
8. I. Chatzigiannakis, T. Dimitriou, M. Mavronicolas, S. Nikoletseas, P. Spirakis. "A Comperative Study of Protocols for Efficient Data Propagation in Smart Dust Networks", PPL. Vol. 13, No 4, pp 615-627, 2003. Also EUROPAR 2003, Distinguished Paper.
9. C. Papadimitriou, D. Ratajczak. "On a Conjecture Related to Geometric Routing", This volume.
10. A. Rao, C. Papadimitriou, S. Shenker, I. Stoika. "Geographic Routing without Location Information", Proc. 9th International Conference on Mobile Computing and Networking, 2003, pp. 96-108, ACM Press.
11. G. S. Lauer. "Packet Radio Routing", Ch. 11 of Routing in Communication Nets, M. Streenstrup (ed.), pp. 351-396, Prentice Hall, 1995.
12. L. M. Kirousis, E. Kranakis, D. Kriznac, A. Pelc. "Power Consumption in Packet Radio Networks", Theoretical Computer Science 243, 289-305, 2000.
13. J. Diaz, P. Spirakis, M. Serna. Document in preparation.

Algorithmic and Foundational Aspects of Sensor Systems*

(Invited Talk)

Paul G. Spirakis

Computer Technology Institute (CTI) and University of Patras,
P. O. Box 1122, 261 10 Patras, Greece
spirakis@cti.gr

Abstract. We discuss here some algorithmic and complexity-theoretic problems motivated by the new technology advances in sensor networks and smart dust. We feel that the field of algorithms has a nice and sizeable intersection with the field of design and control of sensor networks. New measures of efficiency and quality of such algorithms are discussed. Some clear directions for future research are highlighted.

1 Introduction

Sensor networks employ a set of a vast number of very small, fully autonomous computing and communication devices. These devices (the network's *nodes*) should co-operate in order to accomplish a *global* task.

When the devices are ultra-small (abstracted to *points*) then the name *smart dust* (or *smart dust cloud*) is used. We use the term *particles* then for nodes.

The set of nodes of such a network is usually assumed to take *ad hoc* positions in e.g. a surface (2 dimensions) or in an area in the 3-dimensional space. The *area* (or space) *of deployment* of the sensors may include *obstacles* (or *lakes*) i.e. subareas where no sensing device can be found.

Sensor nets differ from general ad-hoc nets with wireless communication in the sense that local resources of each node (such as available energy, storage, communication/computation capability, reliability) are seriously constrained. (See [1], [2] for a nice survey.) In most sensor networks there are also one (or more) *base stations* to which data collected by sensor nodes is relayed for uploading to external systems.

Although the technology of construction of individual cheap sensors is quite advanced today, (e.g. there are sensors of less than a cubic *cm* size, that can even sense a perfume's smell [3]), the corresponding system aspects research is still at its infancy. The general question of "what can a sensors net do/not do globally" is quite open and it is a challenge for the algorithmic thought. Abstract models

---

* This work has been partially supported by the IST Programme of the European Union under contract number IST-2001-33116 (**FLAGS**) and within the 6th Framework Programme under contract 001907 (DELIS).

S. Nikoletseas and J. Rolim (Eds.): ALGOSENSORS 2004, LNCS 3121, pp. 3–8, 2004.

coordinates. Second, the precise coordinates may be disadvantageous as they do not account for obstructions or other topological properties of the network. To address these concerns, Rao et al. [3] recently proposed a scheme in which the nodes first decide on fictitious *virtual coordinates*, and then apply greedy routing based on those. The coordinates are found by a distributed version of the *rubber band* algorithm originally due to Tutte [4] and used often in graph theory [5]. It was noted, on the basis of extensive experimentation, that this approach makes greedy routing much more reliable (in section 3 of this paper we present experimental results on a slight variant of that scheme that has even better performance). However, despite the solid grounding of the ideas in [3] in geometric graph theory, no theoretical results and guarantees are known for such schemes. *The present paper is an attempt to fill this gap: we use sophisticated ideas from geometric graph theory in order to prove the existence of sound virtual coordinate routing schemes.*

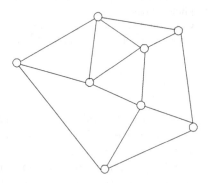

Fig. 1. A *greedy* embedding.

The focal point of this paper is a novel conjecture in geometric graph theory that is elegant, plausible, has important consequences, and seems to be deep. Consider the embedded graph shown in Figure 1. It has the following property: given any nodes s and t, there is a neighbor of s that is closer (in the sense of Euclidean distance) to t than s is. We call such an embedding a *greedy embedding*. We conjecture that *any planar, 3-connected graph has a greedy embedding*. Since every such graph has a convex planar embedding [4] (in which all faces are convex), they are natural candidates for our conjecture, even though there are certainly other graphs with greedy embeddings (for example any graph with a hamiltonian path has a greedy embedding on a straight line). Furthermore, since adding edges only improves the embeddability of a graph, the conjecture extends immediately to any graph with a 3-connected planar subgraph.

2   The Conjecture

Let G be a graph with nodes embedded in $\Re^k$ and let d denote the k-dimensional Euclidean distance. We say that a path $(v_0, v_1, \ldots, v_m)$ is *distance decreasing* if $d(v_i, v_m) < d(v_{i-1}, v_m)$ for $i = 1, \ldots, m$. We propose the following:

Conjecture 1 (Weak). Any planar, 3-connected graph can be embedded on the plane so that between any two vertices s and t there is a distance-decreasing path from s to t.

We call such an embedding a *greedy embedding*, for reasons which are clear from the following theorem.

Theorem 1. *The following are equivalent forms of conjecture 1:*

1. *Any 3-connected planar graph has an embedding on the plane in which for any two nodes s and t there is a neighbor r of s such that $d(r, t) < d(s, t)$.*
2. *Any 3-connected planar graph has an embedding on the plane in which greedy routing will successfully route packets between any source and destination.*
3. *Any graph that contains a 3-connected planar graph has a greedy embedding.*

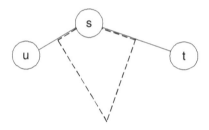

Fig. 2. The forbidden region.

A *convex embedding* of a planar graph is a planar embedding in which all faces, including the external face, are convex. It is known that every 3-connected planar graph has a convex embedding [4] and we additionally conjecture that:

Conjecture 2 (Strong). All 3-connected planar graphs have a greedy convex embedding.

We immediately have the following theorem which is a useful shortcut for testing if an embedding is greedy:

Theorem 2. *A convex embedding of a 3-connected planar graph is greedy if and only if for any obtuse angle about a face (formed by nodes u, s and t), the intersection of the half-planes defined by su (containing t), st (containing u), and the perpendicular bisectors of the segments su and sv (containing s) contain no other vertex of the graph.*

This is illustrated in Figure 2.

Some Counterexamples

We have obtained a family of counterexamples based on a simple lemma.

Lemma 1. *In a greedy embedding, for any node s, s must have an edge to the closest node u in the embedding.*

Proof. Otherwise, u has no neighbor that is closer to s than itself.

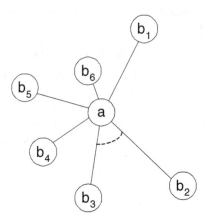

Fig. 3. $K_{1,6}$ has no greedy embedding.

We may therefore construct, for any $k > 0$, a k-connected graph that has no greedy embedding.

Proposition 1. $K_{k,5k+1}$ *has no greedy embedding for $k > 0$.*

Proof. Let A denote the set of nodes in the partition of size k and B denote the remaining nodes. In any embedding, for each element of B we identify which element of A is closest to it. By the pigeonhole principle, there must be 6 nodes $b_1, \ldots, b_6 \in B$ that have some $a \in A$ as the closest element of A. Consider the angles formed by $b_i a b_{i+1}$ (mod 6), as illustrated in Figure 3. At least one of these angles must be no greater than $\pi/3$. Suppose this angle is between b_i and b_{i+1}. By the law of sines, one of the edges ab_i or ab_{i+1} must be no shorter than $b_i b_{i+1}$. This means that one of b_i or b_{i+1} has no edge to the node that is closest to it in the embedding, which by lemma 1 implies that the embedding is not greedy.

These counterexamples imply that the hypotheses of the conjecture are necessary, in that there exist counterexamples that are planar but not 3-connected ($K_{2,11}$), or 3-connected but not planar ($K_{3,16}$); also, they show that high-connectivity alone does not guarantee a greedy embedding.

In our quest to obtain 3-connected planar counterexamples, we have investigated other types of graphs, such as the graph obtained by tiling the plane with

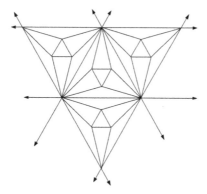

Fig. 4. A graph requiring a non-symmetric embedding.

K_4, or the graph shown in Figure 4. Embedding these graphs require breaking their symmetry and resorting to non-local arguments. However, they do indeed have convex greedy embeddings.

On the Existence of 3-Connected Planar Subgraphs

Assuming the conjecture is true, we may ask which 3-connected graphs have a 3-connected planar subgraph (and hence have a greedy embedding by the conjecture). Are there any 3-connected graphs with no 3-connected planar subgraph?

Such graphs exist, and $K_{3,3}$ is the simplest example: it is both minimally non-planar and minimally 3-connected. However, the following result shows that essentially this is the only counterexample:

Theorem 3. *If a 3-connected graph does not have a $K_{3,3}$ minor, then it has a 3-connected planar subgraph.*

Proof. Suppose the graph has no 3-connected planar subgraph, and consider a maximal planar subgraph G; it must have at least 5 vertices and a cut set of two vertices, a and b. Adding back an edge e to G results in a non-planar graph. By Kuratowski's theorem [] and the hypothesis, $G + e$ has a K_5 minor. Consider these five vertices; at least three of them are neither a nor b. There are three cases, depending on the distribution of these three vertices on the connected components of $G - \{a, b\}$. If they are all on the same component, then G has a K_5 minor and is nonplanar, contradiction. And if they are on two or three different components, they cannot be fully connected to each other via the single edge e.

3   Polyhedral Routing

In this section we consider a "greedy" form of routing on 3-connected planar graphs, where the coordinates are 3-dimensional, and the distance between two points corresponds to their dot product.

Steinitz showed in 1922 that every 3-connected planar graph is the edge graph of a 3-dimensional convex polytope [6]. Such representations are by no means unique, and further work has shown that such a polytope exists even under the constraint that all edges must be tangent to a circle [7]. A crucial property of a convex polytope, employed in the simplex method, is that each vertex of the skeleton has a supporting hyperplane that is tangent to that vertex but which does not intersect the polytope. Furthermore, every other vertex has a neighbor which is closer (in perpendicular distance) to that hyperplane. If the polytope has all edges tangent to a circle centered at the origin (as the Koebe-Andre'ev representation does), then the 3-dimensional coordinates of the corners of the polytope also serve as the normal vector of the supporting hyperplane, as illustrated in Figure 5.

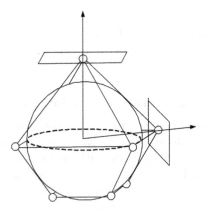

Fig. 5. Koebe-Andre'ev Steinitz representation with supporting hyperplanes.

This implies the following greedy algorithm. Given a 3-connected planar graph G, calculate a Koebe-Andre'ev embedding in 3-dimensions about a circle centered at the origin. For a node v, let $p(v)$ denote its 3-dimensional coordinate in the embedding. When routing to t, s will forward a packet to the neighbor v such that $p(v) \cdot p(t)$ is maximized.

By the convexity of the polytope and the fact that $p(v)$ defines a supporting hyperplane for v, there will always be a neighbor v of s such that $p(v) \cdot p(t) > p(s) \cdot p(t)$. Since $p(v) \cdot p(t)$ is maximized when $v = t$, the above routing algorithm must always make progress until it terminates at the desired destination.

Experiments

The implementation of the above algorithm in a practical setting is confounded by two factors. First, in practical settings, the connection graph need not necessarily be 3-connected or planar. Second, the construction of the Koebe-Andre'ev embedding is not easily obtainable in a distributed fashion [8].

Instead, we consider a variant of the *rubber band* algorithm [3] in which the boundary nodes are evenly spaced on the equator of a unit sphere, and the positions of the remaining nodes are determined by a physical simulation assuming that all edges are equally strong rubber bands that are stretched along the surface of the sphere. For routing, we use the dot-product distance rather than Euclidean distance.

For our simulation, we began with a random 3-connected geometric graph [9] G of 100 nodes placed within a unit square region and with radius of transmission r. We chose three random nodes as boundary nodes (note these need not be on a single face even if G were planar) and then ran both the Euclidean rubber band algorithm in 2-dimensions, and the spherical rubber band algorithm in 3-dimensions. For varying choices of r, we determine the average percentage (over 100 trials) of paths that run into a void and do not terminate at the correct destination.

The results of Figure 3 show that the spherical version of the embedding is a slight improvement over the classical rubber band algorithm under various radii of transmission (increasing r increases the connectivity of the graph). For small values of r, the virtual coordinates of the Euclidean and spherical embeddings perform better than the actual geographic coordinates, an effect which was also noted by Rao et al. [3]. For larger values of r, the graph has larger cliques and becomes less planar. This results in decreased performance of the embeddings as r is increased, and decreased disparity in the peformance of the two embeddings.

Radius (r)	Original	Euclidean	Spherical
0.27	2.347	1.665	1.368
0.28	1.932	1.526	1.253
0.3	1.508	1.486	1.240
0.4	0.165	0.764	0.683
0.5	0.004	0.468	0.469

Fig. 6. The average percentage of paths which are *not* successfully routed.

4   Face Routing

When greedy routing fails, most existing geographic routing algorithms resort to *face routing:* they circumnavigate a face (rather, what appears to the protocol to be a face) until greedy routing can resume [1,3]; such algorithms are either quite complicated, or are not guaranteed to work. In this section we point out that, using a recent strong planar embedding result, we can show that any 3-connected planar graph can be embedded so that a simple and rigorous form of face routing works.

The following statement is equivalent to the existence of the Koebe-Andre'ev Steinitz representation, but restated in a simplified way for our purposes:

Theorem 4. (Koebe, Andre'ev, Thurston) [7]: *Any 3-connected planar graph and its dual can be simultaneously embedded on the plane so that each face is a convex polygon with an inscribed circle whose center coincides with the vertex of the dual corresponding to the face, and so that edges are perpendicular to their dual edges.*

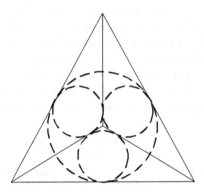

Fig. 7. Koebe-Andre'ev planar representation with inscribed circles.

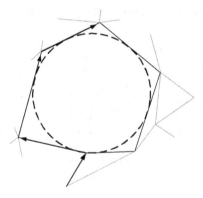

Fig. 8. Face routing on a Koebe-Andre'ev embedding.

Figure 7 illustrates an example of this embedding. This result gives rise to the following rigorous face routing protocol: each vertex knows, besides the coordinates of its neighbors, the coordinates of all dual vertices representing the faces adjacent to it; each such face points to the neighbor vertex that comes next in clockwise order. When greedy routing fails, then the vertex employs face routing: it selects the face (dual vertex) that is closest to the destination, and sends the packet clockwise around that face. When a vertex is found with a neighbor that is closer to the destination than the vertex that began the face routing, greedy routing is re-established. This process is illustrated in Figure 8. We omit the proof of the following:

Theorem 5. *Any graph containing a 3-connected planar graph can be embedded on the plane so that the above face routing protocol always succeeds.*

5   Discussion

We have thus far been unable to prove the conjecture, even for the special case when the graph is a maximally planar graph (all faces are triangles). One particularly attractive property of 3-connected planar graphs is a lemma of Thomasson stating that all 3-connected planar graphs of more than 4 vertices have an edge that can be contracted to yield another 3-connected planar graph. We initially wished to inductively construct the embedding based on this lemma (and some of its variants). However, all efforts in this direction have failed. While we have confidence in the truth of the conjecture, we suspect that its proof is non-trivial.

Acknowledgments. We would like to thank Janos Pach and Nati Linial for their ideas and insights regarding the conjecture.

References

1. Kuhn, F., Wattenhofer, R., Zhang, Y., Zollinger, A.: Geometric Ad-Hoc Routing: Of Theory and Practice". In: Proc. 22nd ACM Int. Symposium on the Principles of Distributed Computing (PODC). (2003)
2. Karp, B., Kung, H.T.: GPSR: greedy perimeter stateless routing for wireless networks. In: Mobile Computing and Networking. (2000) 243–254
3. Rao, A., Papadimitriou, C., Shenker, S., Stoica, I.: Geographic routing without location information. In: Proceedings of the 9th annual international conference on Mobile computing and networking, ACM Press (2003) 96–108
4. Tutte, W.T.: Convex representations of graphs. Proceedings London Math. Society **10** (1960) 304–320
5. Linial, N., Lovasz, L., Wigderson, A.: Rubber bands, convex embeddings and graph connectivity. Combinatorica **8** (1988) 91–102
6. Ziegler, G.M.: Lectures on Polytopes. Springer-Verlag, Berlin (1995)
7. Thurston, W.: Three-dimensional Geometry and Topology. Number 35 in Princeton Mathematical Series. Princeton University Press, Princeton, NJ (1997)
8. Lovasz, L.: Steinitz representations of polyhedra and the colin de verdiere number. Journal of Combinatorial Theory **B** (2001) 223–236
9. Penrose, M.: Random Geometric Graphs. Oxford University Press (2003)

WiseMAC: An Ultra Low Power MAC Protocol for Multi-hop Wireless Sensor Networks

Amre El-Hoiydi and Jean-Dominique Decotignie

CSEM, Swiss Center for Electronics and Microtechnology, Inc,
Rue Jaquet-Droz 1,
2007 Neuchâtel, Switzerland
{aeh, jdd}@csem.ch

Abstract. WiseMAC is a medium access control protocol designed for wireless sensor networks. This protocol is based on non-persistent CSMA and uses the preamble sampling technique to minimize the power consumed when listening to an idle medium. The novelty in this protocol consists in exploiting the knowledge of the sampling schedule of one's direct neighbors to use a wake-up preamble of minimized size. This scheme allows not only to reduce the transmit and the receive power consumption, but also brings a drastic reduction of the energy wasted due to overhearing. WiseMAC requires no set-up signalling, no network-wide synchronization and is adaptive to the traffic load. It presents an ultra-low power consumption in low traffic conditions and a high energy efficiency in high traffic conditions. The performance of the WiseMAC protocol is evaluated using simulations and mathematical analysis, and compared with S-MAC, T-MAC, CSMA/CA and an ideal protocol.[1]

1 Introduction

A wireless sensor network [1] is composed of numerous nodes distributed over an area to collect information. The sensor nodes communicate among them through the wireless channel to self-organize into a multi-hop network and forward the collected data towards one or more sinks. Because of the difficulty to recharge or replace the battery of each node in a sensor network, the energy efficiency of the system is a major issue. A meticulous design of the medium access control (MAC) protocol is key to reach a low power consumption.

Sensor networks are usually meant for the acquisition of data, either periodically and/or based on events. Applications include for example agriculture monitoring, environmental control in building or alarm systems. In most scenarios, it is anticipated that a sensor node will be idle most of the time. For example, in some agriculture monitoring application, measurements reported at

---

[1] The work presented in this paper was supported in part by the National Competence Center in Research on Mobile Information and Communication Systems (NCCR-MICS), a center supported by the Swiss National Science Foundation under grant number 5005-67322.

an interval of 1 hour may be satisfactory. In an alarm system, one would expect a periodic traffic from the alarms, with a period of e.g. 1 minute, to inform the sink (a central controller) that they are still in operation. From time to time, detected events can cause a large traffic to flow from the alarms to the sink. In alarm applications, a low transmission delay can be a very important additional requirement, beside the low power consumption requirement.

An energy efficient wireless MAC protocol must minimize the four sources of energy waste [14]: idle listening, overhearing, collisions and protocol overhead. Idle listening refers to the active listening to an idle channel, waiting for a potential packet to arrive. Overhearing refers to the reception of a packet, or of part of a packet, that is destined to another node. Collisions should of course be avoided as retransmissions cost energy. Finally, protocol overhead refers to the packet headers and the signalling required by the protocol in addition to the transmission of data payloads.

With a conventional CSMA protocol, nodes listen to the radio channel whenever they do not transmit. As the power consumption of a transceiver in receive mode is far from being negligible, idle listening becomes clearly the main source of energy waste in scenarios where the channel is idle most of the time. Low power MAC protocols must use techniques to mitigate idle listening. Overhearing must however not be underestimated. If the idle listening problem is efficiently addressed by a MAC protocol, the following important source of energy waste becomes overhearing, especially in dense ad-hoc networks.

This paper presents WiseMAC (*Wi*reless *Se*nsor *MAC*), a novel medium access control protocol for multi-hop wireless sensor networks. This protocol has been briefly outlined in a poster abstract [2]. In [4], its performances were analyzed for the downlink of an infrastructure wireless network (i.e. a collisionless channel) in very low traffic conditions. The original contribution of this paper is the detailed presentation of the WiseMAC protocol and the study of its performance in a multi-hop wireless sensor network, in comparison with other protocols previously proposed by other authors.

The remaining of the paper is organized as follows: Section 2 presents related work. WiseMAC is described in Section 3. Its performance is analyzed in Section 4, considering both a lattice topology with parallel traffic and a random network topology with traffic collected at a sink. Finally, Section 5 gives concluding remarks.

2   Related Work

Among the contention-less medium access control protocols (time, frequency and code division multiple access), TDMA appears as an appealing candidate for a low power MAC protocol, as it causes neither overhearing nor collisions. Sensor nodes may sleep in-between assigned communication slots. In the context of sensor networks, it was proposed in [11], together with a TDMA schedule setup algorithm and signalling protocol. TDMA has several weaknesses when used for wireless sensor networks. First, it is only energy efficient when transporting periodic traffic. When the traffic is sporadic, many communication slots are empty and energy is thereby wasted. Secondly, the signalling protocol required to setup and maintain the Spatial-TDMA schedule, and to maintain the network synchronization may be very complex and costly both in computational and energetic resources.

Schurgers *et al.* have proposed STEM [10]. This protocol uses two channels: one paging channel and one traffic channel. Most of the time, the network is expected to be in the monitoring state, and only the paging channel is used. In case of an alarm, for example, a path on the data channel is opened throughout the network, where communication occurs using regular wireless protocols (not low power). In STEM, the paging channel is implemented at the receiver side by regularly listening to the channel during the time needed to receive a paging packet. A transmitter that wants to page one of its neighbor will repeat a paging packet containing the destination address, until a reply is received. This protocol provides a low power consumption in the absence of traffic, the paging channel consuming little energy. A traffic burst resulting from an event detection can be transported in an energy efficient way using a classical protocol. The weakness of this protocol is mainly its inefficiency to transport small amount of periodic or sporadic traffic.

Ye *et al.* have proposed S-MAC (Sensor-MAC), which provides a low duty cycle without the need to precisely synchronize the sensor network [14]. This protocol defines sleep intervals, in which all the nodes of the network sleep, and active intervals, in which communication can occur. Because listen intervals are relatively large, only a loose synchronization is required among neighboring nodes. During the listen interval, sensor nodes having traffic to send attempt to reserve the medium and signal to the destination to remain awake for the data transmission using the request-to-send/clear-to-send packet exchange mechanism. With S-MAC, one must select the frame duration (i.e. the total of the listen and sleep intervals), as a trade-off between the average power consumption and the transmission delay. S-MAC exploits the concept of fragmentation to transmit large messages in an energy efficient way.

Van Dam *et al.* have proposed T-MAC (Timeout-MAC) [12]. T-MAC is an improvement of S-MAC. In the T-MAC protocol, the length of the active period is dynamically adapted to the traffic, using a timeout. The active period is ended whenever physical and virtual carrier sensing find the channel idle for the duration of the time-out. A similar idea was also and independently proposed by Ye *et al.* in [15].

S-MAC and T-MAC can currently be seen as benchmarks in the field of contention MAC protocols for sensor networks. WiseMAC will be compared to them.

Rabaey *et al.* have proposed a hardware based solution to wake-up a destination node [9]. They suggest the use of a separate super-low-power wake-up radio that will switch the main radio on at the start of the data packet. This solution is of great interest as it would provide small hop latencies. The wake-up preamble being short, this method would also preserve the channel capacity. The development of such a super-low-power wake-up radio consuming only a few tens of μW being still a challenge, solutions using conventional radio transceivers remain of interest. If such a wake-up radio becomes available, it may also be envisaged to use it in combination with low power MAC protocols, to reduce even further the power consumption.

3   WiseMAC

WiseMAC is based on the preamble sampling technique [3]. This technique consists in regularly sampling the medium to check for activity. By *sampling the medium*, we mean listening to the radio channel for a short duration, e.g. the duration of a modulation symbol. All sensor nodes in a network sample the medium with the same constant period T_W. Their relative sampling schedule offsets are independent. If the medium is found busy, a sensor node continues to listen until a data frame is received or until the medium becomes idle again. At the transmitter, a wake-up preamble of size equal to the sampling period is added in front of every data frame to ensure that the receiver will be awake when the data portion of the packet arrives. This technique provides a very low power consumption when the channel is idle. The disadvantages of this protocol are that the (long) wake-up preambles cause a throughput limitation and a large power consumption overhead in transmission and reception. The overhead in reception is not only bared by the intended destination, but also by all other nodes overhearing the transmission. The WiseMAC protocol aims at reducing the length of this costly wake-up preamble.

The novel idea introduced by WiseMAC consists in learning the sampling schedule of one's direct neighbors to use a wake-up preamble of minimized size. It will be shown that this simple idea provides a significant improvement compared to the basic preamble sampling protocol, as well as to S-MAC and T-MAC. The basic algorithm of WiseMAC works as follows:

Because the wireless medium is error prone, a link level acknowledgement scheme is required to recover from packet losses. The WiseMAC ACK packets are not only used to carry the acknowledgement for a received data packet, but also to inform the other party of the remaining time until the next sampling time. In this way, a node can keep a table of sampling time offsets of all its usual destinations up-to-date. Using this information, a node transmits a packet just at the right time, with a wake-up preamble of minimized size, as illustrated in Fig. 1.

The duration of the wake-up preamble must cover the potential clock drift between the clock at the source and at the destination. This drift is proportional to the time since the last re-synchronization (i.e. the last time an acknowledgement was received). Let θ be the frequency tolerance of the time-base quartz and L the interval between communications. As shown below, the required duration of the wake-up preamble is given by

$$T_P = \min(4\theta L, T_W) . \tag{1}$$

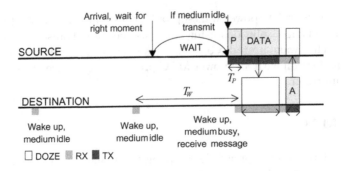

Fig. 1. WiseMAC

Expression (1) can be derived as follows: Assume that both the source and the destination are equipped with a clock based on a quartz with a tolerance of $\pm\theta$. Assume that the source has received fresh timing information from some sensor node at time 0, and that it wants to send a packet to this sensor node at the sampling time L. If the destination sensor node quartz has a real frequency of $f(1+\theta)$ instead of f, its clock will have an advance of θL at time L. It is hence needed to start the preamble transmission θL in advance. Because the clock of the source might be late, the source must target a time $2\theta L$ in advance to L. Because the clock of the source might be early, and the clock of the destination late, the duration of the wake-up preamble must be of $4\theta L$. If L is very large, $4\theta L$ may be larger that the sampling period T_W. In those cases, a preamble of length T_W is used. We thus obtain $T_P = \min(4\theta L, T_W)$.

The transmission will be started at time $L - T_P/2$, to center the wake-up preamble on the expected scheduled sampling. If the medium is sensed busy at the scheduled transmission time, the attempt is deferred using non-persistent CSMA.

The first communication between two nodes will always be done using a long wake-up preamble (of length T_W). Once some timing information is acquired, a wake-up preamble of reduced size can be used. The length of the wake-up preamble being proportional to the interval L between communications, it will be small when the traffic is high. This important property makes the WiseMAC protocol adaptive to the traffic. The per-packet overhead decreases with increasing traffic. In low traffic conditions, the per-packet overhead is high, but the average power consumption caused by this overhead is low.

Overhearing is naturally mitigated when the traffic is high, thanks to the combined use of the preamble sampling technique and the minimization of the wake-up preamble duration. As already mentioned, sensor nodes are not synchronized among themselves. Their relative sampling schedule offsets are independent. In high traffic conditions, the duration of the wake-up preamble being smaller than the sampling period, short transmission are likely to fall in between sampling instants of potential overhearers.

The synchronization mechanism of WiseMAC can introduce a risk of systematic collision. Indeed, in a sensor network, a tree network topology with a number of sensors sending data through a multi-hop network to a sink often occurs. In this situation, many nodes are operating as relays along the path towards the sink. If a number of sensor nodes try to send a data packet to the same relay, at the same scheduled sampling time and with wake-up preambles of approximately identical size, there are high probabilities to obtain a collision. To mitigate such collisions, it is necessary to add a medium reservation preamble of randomized length in front of the wake-up preamble. The sensor node that has picked the longest medium reservation preamble will start its transmission sooner, and thereby reserve the medium.

A very important detail of the WiseMAC protocol, which is also found in the IEEE 802.11 power save protocol, is the presence of a *more* bit in the header of data packets. When this bit is set to 1, it indicates that more data packets destined to the same sensor node are waiting in the buffer of the transmitting node. When a data packet is received with the more bit set, the receiving sensor node continues to listen after having sent the acknowledgement. The sender will transmit the following packet right after having received the acknowledgement. This scheme permits to use a sampling period that is larger than the average interval between the arrivals for a given node. It also permits to reduce the end-to-end delay, especially in the event of traffic bursts.

The *more* bit scheme provides the same functionality as the fragmentation scheme used in S-MAC [14]. An application just needs to segment a large message into smaller packets to obtain the fragmentation behavior. However, the *more* bit scheme is more flexible. Packets that do not belong to the same message but that need to be sent to the same destination will be grouped when using the *more* bit, while they would be sent individually with the fragmentation scheme.

4   Performance Analysis

In this section, the performance of the WiseMAC protocol will be analyzed through simulation and mathematics. Comparisons will be made with the basic preamble sampling protocol [3] (called hereafter BPS), S-MAC, T-MAC, CSMA/CA and an ideal protocol. The purpose of the ideal protocol is to provide a target benchmark for implementable protocols. With the ideal protocol, a node 'magically' knows some time in advance that it has to power-on the receiver to receive a packet. The transceiver is only consuming energy to transmit and receive packets. There is absolutely no idle listening, overhearing nor collision overhead. Real protocols will always consume some energy to implement the wake-up scheme. Their comparison with the ideal protocol will indicate their overhead.

The performance of the WiseMAC protocol has been studied using both a regular lattice topology with traffic flowing in parallel and a typical sensor network topology with randomly positioned sensors forwarding data towards a sink.

The interest of a lattice topology [7] with traffic flowing in parallel is that it allows exploring the behavior of a MAC protocol without inserting aspects linked to routing, load balancing and traffic aggregation. The regularity of the topology allows deriving mathematical expressions to approximate the power consumption. After having introduced the system parameters, we will start the analysis considering the lattice topology in Section 4.2. The random network topology will be addressed in Section 4.3.

4.1 System Parameters

WiseMAC, BPS, S-MAC, T-MAC and CSMA/CA have been modeled on the GloMoSim platform [16]. The radio layer of GloMoSim has been extended to include a DOZE state and to mimic precisely the transition delays of the modeled radio transceiver. The protocols performance is computed using the parameters of the low power FSK radio transceiver developed at CSEM within the WiseNET project [8]. The power consumed in DOZE, RX and TX states is respectively $P_Z = 5 \ \mu W$, $P_R = 1.8$ mW and $P_T = 27$ mW. The time needed to turn-on the transceiver into the RX or TX state is equal to 0.8 ms and the time to turn it around between these states is equal to 0.4 ms. A bit rate of 25 kbps is assumed. Relatively similar performance curves are obtained with other low power transceivers, such as the RFM TR1000 and Chipcon CC1000 used on Berkeley's mica and mica2 motes [6].

With the WiseMAC protocol, one must select the sampling period, as a trade-off between the hop delay and the average power consumed by the sampling activity. T_W should be chosen large enough such that only a fraction of the power budget is consumed by the sampling activity. The larger the value of T_W, the smaller the power consumption of the sampling activity and the larger the hop delay. However, in terms of lifetime and when using a leaking battery, it doesn't pay to have a power consumption for the sampling activity that is negligible compared to the leaked power. With the WiseNET transceiver and when using an AA alkaline battery with about 30 μW leakage power, a good value for the sampling period was found to be $T_W = 200$ ms . Using a larger value for T_W increases linearly the delay without increasing much the lifetime.

With S-MAC and T-MAC, 3 different frame durations will be used such as to provide a duty cycle of 1, 5, and 10% in the absence of traffic (10% is the default duty cycle in the implementation of S-MAC on Berkeley's motes [13]). This corresponds to a frame duration of respectively 2000, 400 and 200 ms in the case of S-MAC and 2400, 480 and 240 ms in the case of T-MAC.

4.2 Lattice Network

Topology. In this section, we consider a lattice network topology as illustrated in Fig. 2. A separation of 40 meters between nodes is assumed. As in [12], the number of neighbors in range (the node degree) is chosen to be $N = 8$. This number has been chosen small to limit the local traffic but large enough to provide a well connected topology in a random plane network with the same

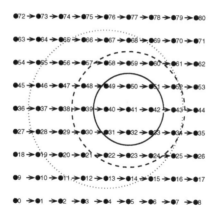

Fig. 2. Lattice network topology

node density (see the percolation theory [5]). Hence, obtained results are also applicable to random plane ad-hoc networks of equal degree.

Traffic. Poisson traffic is generated by nodes on the left (0,9, ..., 72) and transmitted in multi-hop fashion towards nodes on the right (8, 17, .., 80). The rate λ of the packet generation is constant throughout a simulation. As long a no packets are dropped due to congestion, every node in the network forwards packets at rate λ. Simulations results are shown for packet generation rates varying between 0.001 and 1 packet per second (inter-arrivals L between 1000 s and 1 s). Such traffic can be expected in a sensor network for example as a result of a regular data acquisition (e.g. temperature monitoring) or the periodic transmission of alive messages by alarm sensors. The power consumption calculations are done for node number 40. As was shown in [7], the behavior of this central node is approaching the behavior of a forwarding node in a very large network. Having fewer neighbors, nodes on the sides of the network will suffer less from overhearing, collisions and backoff.

The size of the data portion of the packets is chosen to be of 48 bytes. The MAC layer overhead is of 8 bytes. The resulting frame size is of 56 bytes, and its transmission duration at 25 kbps is of 18 ms. An acknowledgement packet has a length of 12 bytes and a duration of 4 ms. Note that, in an optimized implementation, smaller overheads could be used.

Receive, Interference and Carrier Sense Ranges. The sensitivity of a transceiver is the minimum energy level at which a signal can be demodulated at some given bit error rate (typically 10^{-3}). In WiseMAC, we have chosen to use a receive threshold that is well above the sensitivity, in order to avoid useless wake-ups caused by noise or by weak signals, and wake up only when this is really worth it. Here, the lower power consumption is traded against a potential transmission range extension. With the two-ray propagation model, the expected reception range for the WiseNET transceiver is 62 meters, as illustrated in Fig. 2 by the solid circle around node 41. All nodes located within this circle may transmit a packet successfully to node 41.

The radio layer model used in the simulation is based on a required SNR. If the sum of all interfering signals is, at any moment during the packet reception, above the packet signal power minus the required SNR, then the packet is declared lost. A SNR of 14 dB is chosen in order to declare a packet received without errors. If the source of the transmission is 40 meters away (node 40), this results in an interference range of 100 meters, as illustrated by the dashed circle around the node number 41.

The dotted circle around node 40 represents the sensitivity range, i.e. the range up to which a transmission is detected, and the medium is declared non-idle. A transmission is initiated by node 40 only if the medium is found idle, corresponding to the situation where none of the 37 nodes located within the dotted circle is transmitting. A reception is attempted only if the data is received at a power above the receive threshold. Having a carrier sensing range larger than the interference range yields a protection against the hidden node effect: all nodes located in the interference range of node 41 are within the sensitivity range of node 40. However, this very interesting property is gained at the cost of a reduced overall maximum throughput, as only one of 37 nodes may transmit at the same time. In a sensor network, this maximum throughput reduction can however be acceptable.

Power Consumption and Hop Delay. Fig. 3 presents the average power consumption (upper plot) and the average hop delay (lower plot) as a function of the traffic. Power consumption results are collected from simulations by recording the time spent by the radio transceiver of node 40 in the states DOZE, RX and TX. The average power is computed as $\sum_{i \in States} r_i P_i$, where r_i is the proportion of time spent in state i and P_i is the power consumed in that state. This average power corresponds to the task of forwarding one packet every L seconds. Power consumption values resulting from simulation have been verified through a theoretical analysis. The dashed lines in Fig. 3 show the theoretical power consumption of WiseMAC, BPS, S-MAC and the ideal protocol (The mathematical expressions and their derivation could unfortunately not be included in this paper for space limitation reasons). The hop delay has been obtained from the simulation by dividing the average end-to-end transmission delay between nodes 36 and 44 by the number of hops (8). The curves are plotted up to an injection rate that causes more than 5% packet loss.

It can be seen in Fig. 3 that WiseMAC provides both a low power consumption and a relatively low hop delay, while the other protocols provide only one or the other.

The comparison between WiseMAC and BPS shows the gain brought by the minimization of the length of the wake-up preamble. For $L = 100$, WiseMAC consumes 3.8 times less power than BPS. For higher traffic, the advantage is even

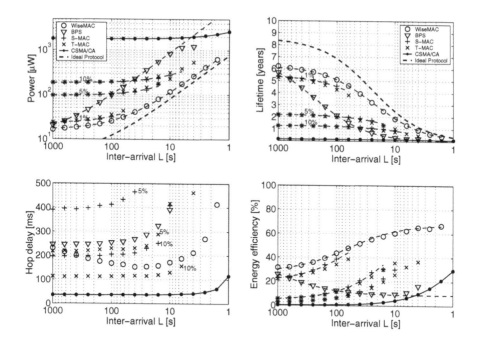

Fig. 3. Average power consumption and hop delay

Fig. 4. Lifetime using a single AA alkaline battery and energy efficiency

greater. As it does not mitigate idle listening, CSMA/CA consumes a minimum of $P_R = 1.8$ mW, which is 70 times more than WiseMAC for $L = 100$. S-MAC and T-MAC consume at least a fraction of P_R, corresponding to the selected duty cycle ($0.1 \cdot P_R = 180$ μW with 10% duty cycle). The power consumption of S-MAC and T-MAC increases with increasing traffic. For $L = 100$, WiseMAC consumes 7 times less than S-MAC or T-MAC at 10% duty cycle. When used at 1 % duty cycle, S-MAC and T-MAC are closer to WiseMAC in terms of power consumption, but are penalized by a hop delay respectively 10 and 6 times higher than what is provided by WiseMAC.

The hop delay with S-MAC is approximately equal to the frame duration. With T-MAC, it is equal to half the frame duration. These simulation results are in line with the theoretical latency analysis presented in [15]. With BPS, the hop delay is at least equal to the sampling period (as this is also the size of the wake-up preamble). With WiseMAC, the hop delay can become smaller than the sampling period due to the use of short wake-up preambles.

As previously mentioned, simulation results are plotted up to an injection rate that causes more than 5% packet loss. Knowing this, one can notice in Fig 3 (upper plot) that WiseMAC can provide a higher throughput than S-MAC or T-MAC.

Lifetime and Energy Efficiency. The gain brought by a lower power consumption in low traffic conditions is best visible when looking at the lifetime that can be reached with an AA alkaline battery. We use here the battery model introduced in [3]. Fig. 4 (upper part) shows that a lifetime of 5 years can be achieved with WiseMAC when forwarding packets at a rate of 1 every 100 seconds. With BPS, only 2 years are reached. With S-MAC-10% and T-MAC-10%, a little more than 1 year is reached. If the sensor network is operated constantly under a high traffic load (1 packet per second), the lifetime will be very limited with any protocol, even the ideal one. This shows that, if several years of lifetime is a requirement, then high traffic periods should be kept rare.

A meaningful metric for the comparison of low power MAC protocols, especially in high traffic conditions, is their energy efficiency. We define the energy efficiency of a MAC protocol as the ratio between the average power consumed by the ideal protocol and the average power consumed by the protocol of interest. The resulting energy efficiency curves for the different protocols are shown in the lower part of Fig. 4. It can be seen that all protocols have a relatively low energy efficiency in low traffic conditions. Each protocol is associated with a constant minimum power consumption, even in the absence of traffic. With WiseMAC and BPS, this minimum overhead is the sampling activity. With S-MAC and T-MAC, it is the cost of periodically listening to the channel. When the traffic increases, the energy efficiency increases, as this overhead is shared among more packets. With WiseMAC and BPS, there is additionally the overhead of the wake-up preamble. With WiseMAC, the length of the wake-up preamble becomes small when the traffic increases, while it remains constant with BPS. WiseMAC reaches an energy efficiency over 60%. The energy efficiency of the other protocols remains below 40%.

4.3   Random Network

Topology and Traffic. Wireless sensor network are often foreseen to operate in a random multi-hop network topology, where sensors forward data to one or more sinks. Such a topology, as illustrated in Fig. 5, will be considered in this section. The network is composed of 90 sensor nodes, spread over an area of 300×300 meters. Traffic is generated by the 10 black nodes and relayed by the white nodes towards the sink, located on the lower right corner. Routing is precomputed using Dijkstra's algorithm. The resulting minimum hop routing tree is represented by black lines. The remaining and unused links are represented by gray lines.

The following three experiments have been made:

Idle. No traffic is generated. The simulation is run for 4000 (simulated) seconds (about 1 hour).

Distributed traffic. The 10 black nodes generate periodically, with a period of 400 s, a packet of 48 bytes (56 bytes with MAC header). The first node starts at time 0, the second at time 40, ..., the last one at time 360. Traffic is thus distributed over time. As long as the end-to-end delay remains below

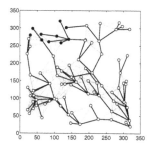

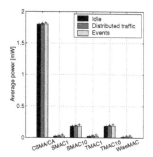

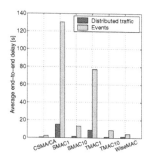

Fig. 5. Random network topology

Fig. 6. Average power consumption

Fig. 7. Average end-to-end delay

40 seconds (which will be the case in this experiment), only one packet is in the network at any time. The simulation is run for 4000 seconds. A total of 100 packets is hence generated.

Events. The black nodes generate periodically, with a period of 400 s, a packet of 48 bytes (56 bytes with MAC header). They all start at the same times 0, 400, 800, ..., 3600 s. This generate periodically a burst of traffic. Again, the simulation is run for 4000 seconds and a total of 100 packets is generated.

The purpose of the *distributed traffic* experiment is to explore the behavior of MAC protocols in low traffic conditions. Such a traffic pattern can be expected in many environmental monitoring applications, such as for the periodic measurement of soil moisture in smart agriculture.

The purpose of the *events* experiment is to explore the behavior of MAC protocols in momentary high traffic conditions. Such a traffic pattern can be expected in alarm systems, such as fire or motion detection sensor networks.

In both experiments, a total 100 packets are forwarded towards the sink. In the *events* experiment, events have been spaced sufficiently such that only 10 packets are in the network at any time. The buffer capacity on each sensor node being of 10 packets, no packets will be lost. Some protocols will require more time to transport the 10 packets than others.

A comparison of the power consumption and delay performances of WiseMAC, S-MAC, T-MAC and CSMA/CA is made in the next sub-section.

Power Consumption and Delay. The bars in Fig. 6 show, for the different experiments and MAC protocols, the average power consumption spent by the nodes. To compute the average power, the total consumed energy is divided by the number of nodes and the simulation time. This average power gives information about the total energy spent in the network. As the lifetime of a network is often bounded by the lifetime of its weakest nodes, it is important to consider also the maximum average power consumed by any node. It is shown as the "+" markers in Fig. 6.

Fig. 7 shows the corresponding average end-to-end transmission delay. This is the average time required by the 100 packets to reach the sink.

The CSMA/CA protocol provides, of course, the lowest average delay for both distributed and events traffic. This is however payed by a power consumption that is much higher than all other protocols. The power consumption of CSMA/CA is lower bounded by the power consumption in receive mode $P_R = 1.8$ mW.

S-MAC-1% and T-MAC-1% provide a low average power consumption, in the order of what is provided by WiseMAC. However, the corresponding delay is very high, while it remains low for WiseMAC. S-MAC-10% and T-MAC-10% are able to provide a relatively low delay, but at the price of a power consumption that is much higher than the one of WiseMAC.

WiseMAC is able to provide a low average transmission delay even in the events experiment. This is made possible by the 'more' bit, which aggregates packets along the path towards the sink and transmit them in bursts.

5   Conclusion

WiseMAC is a contention protocol using the preamble sampling technique to mitigate idle listening. The novel idea introduced by WiseMAC is to minimize the length of the wake-up preamble, exploiting the knowledge of the sampling schedule of one's direct neighbors. Since a node will have only a few direct destinations, a table listing their sampling time offset is manageable even with very limited memory resources.

WiseMAC presents many appealing characteristics. It is scalable as only local synchronization information is used. It is adaptive to the traffic load, providing an ultra low power consumption in low traffic conditions and a high energy efficiency in high traffic conditions. Thank to the 'more' bit, WiseMAC can transport bursty traffic, in addition to sporadic and periodic traffic. This protocol is simple, in the sense that no complex signalling protocol is required. This simplicity can become crucial when implementing WiseMAC on devices with very limited computational resources.

WiseMAC was compared to S-MAC and T-MAC both in a regular lattice topology with traffic flowing in parallel, and in a random network topology with periodic or event traffic flowing towards a sink. When forwarding packets at an interval of 1 packets every 100 seconds, the power consumption of WiseMAC was found to be 25 μW, providing 5 years of lifetime using a single AA alkaline battery. This was seen to be 70 times better than CSMA/CA and 7 times better than S-MAC and T-MAC at a duty cycle of 10%. It was shown that WiseMAC can provide simultaneously a low hop delay and a low power consumption, while S-MAC and T-MAC can only provide one or the other. Finally, it was shown that WiseMAC is able to provide a higher throughput than both S-MAC-10% and T-MAC-10%.

Acknowledgements. The authors would like to thank Erwan Le Roux for having guided us towards the preamble sampling technique.

References

1. I. F. Akyildiz, W. Su, Y. Sankarasubramaniam, and E. Cayirci. Wireless sensor networks: A survey. *IEEE Communications Magazine*, 40(8):102–114, August 2002.
2. A. El-Hoiydi, J.-D. Decotignie, C. Enz, and E. Le Roux. Poster Abstract: WiseMAC, An Ultra Low Power MAC Protocol for the WiseNET Wireless Sensor Network. In *Proc. 1st ACM SenSys Conf.*, pages 302–303, November 2003.
3. Amre El-Hoiydi. Spatial TDMA and CSMA with Preamble Sampling for Low Power Ad Hoc Wireless Sensor Networks. In *Proc. IEEE Int. Conf. on Computers and Communications (ISCC)*, pages 685–692, Taormina, Italy, July 2002.
4. Amre El-Hoiydi, Jean-Dominique Decotignie, and Jean Hernandez. Low Power MAC Protocols for Infrastructure Wireless Sensor Networks. In *Proc. European Wireless (EW'04)*, pages 563–569, Barcelona, Spain, February 2004.
5. E. N. Gilbert. Random Plane Networks. *Journal of the Society of Industrial and Applied Mathematics*, 9(4):533–543, Dec 1961.
6. Jason L. Hill and David E. Culler. Mica: a wireless platform for deeply embedded networks. *IEEE Micro*, 22(6):12–24, Nov.-Dec. 2002.
7. J. Li et al. Capacity of Ad Hoc wireless networks. In *Proc. ACM/IEEE MOBICOM Conf*, pages 61–69, 2001.
8. T. Melly, E. Le Roux, F. Pengg, D. Ruffieux, N. Raemy, F. Giroud, A. Ribordy, and V. Peiris. WiseNET: Design of a Low-Power RF CMOS Receiver Chip for Wireless Applications. *CSEM Scientific and Technical Report*, page 25, 2002.
9. J. Rabaey et al. Picoradio supports ad hoc ultra-low power wireless networking. *IEEE Computer Magazine*, pages 42–48, July 2000.
10. Curt Schurgers, Vlasios Tsiatsis, and Mani B. Srivastava. STEM: Topology Management for Energy Efficient Sensor Networks. In *Proc. IEEE Aerospace Conf.*, volume 3, pages 1099–1108, Mar 2002.
11. K. Sohrabi, J. Gao, V. Ailawadhi, and G.J. Pottie. Protocols for self-organization of a wireless sensor network. *IEEE Personal Communications*, 7(5):16–27, Oct. 2000.
12. Tijs van Dam and Koen Langendoen. An adaptive Energy-Efficient MAC Protocol for Wireless Sensor Networks. In *Proc. ACM SenSys*, pages 171–180, Nov. 2003.
13. Wei Ye, John Heidemann, and Deborah Estrin. A Flexible and Reliable Radio Communication Stack on Motes. Technical report ISI-TR-565, USC/ISI, Sept 2002.
14. Wei Ye, John Heidemann, and Deborah Estrin. An Energy-Efficient MAC Protocol for Wireless Sensor Networks . In *Proc. IEEE INFOCOM Conf.*, 2002.
15. Wei Ye, John Heidemann, and Deborah Estrin. Medium Access Control with Coordinated, Adaptive Sleeping for Wireless Sensor Networks. Technical Report ISI-TR-567, USC/Information Sciences Institute, Jan 2003.
16. X. Zeng, R. Bagrodia, and M. Gerla. GloMoSim: a Library for Parallel Simulation of Large-scale Wireless Networks. In *Proc. Int. Workshop on Parallel and Distributed Simulations*, pages 154–161, May 1998.

On the Computational Complexity of Sensor Network Localization

James Aspnes*, David Goldenberg**, and Yang Richard Yang***

Yale University
Department of Computer Science
New Haven, CT 06520-8285, USA
{aspnes,yry}@cs.yale.edu, david.goldenberg@yale.edu

Abstract. Determining the positions of the sensor nodes in a network is essential to many network functionalities such as routing, coverage and tracking, and event detection. The localization problem for sensor networks is to reconstruct the positions of all of the sensors in a network, given the distances between all pairs of sensors that are within some radius r of each other. In the past few years, many algorithms for solving the localization problem were proposed, without knowing the computational complexity of the problem. In this paper, we show that no polynomial-time algorithm can solve this problem in the worst case, even for sets of distance pairs for which a unique solution exists, unless **RP = NP**. We also discuss the consequences of our result and present open problems.

1 Introduction

Localization is the process by which the positions of the nodes in an ad-hoc network are determined. Knowing the positions of the network nodes is essential because many other network functionalities such as location-dependent computing (e.g., [11, 29]), geographic routing (e.g., [17]), coverage and tracking (e.g., [18]), and event detection depend on location. Although localization can be achieved through manual configuration or by exploiting the Global Positioning System (GPS) [15], neither methodology scales well and both have physical limitations. For example, GPS receivers are costly both in terms of hardware and power requirements. Furthermore, since GPS reception requires line-of-sight between the receiver and the satellites, it may not work well in buildings or in the presence of obstructions such as dense vegetation, foliage, or mountains blocking the direct view to the GPS satellites.

Recently, novel schemes have been proposed to determine the locations of the nodes in a sensor network, e.g., [1, 2, 3, 4, 5, 6, 8, 10, 12, 13, 14, 16, 19, 20, 21, 22, 23, 25, 27, 28, 30]. In these schemes, network nodes measure the distances to

* Supported in part by NSF.
** Supported in part by an NSF Graduate Research Fellowship.
*** Supported in part by NSF.

S. Nikoletseas and J. Rolim (Eds.): ALGOSENSORS 2004, LNCS 3121, pp. 32–44, 2004.
© Springer-Verlag Berlin Heidelberg 2004

their neighbors and then try to compute their locations. Although the designs of the previous schemes have demonstrated clever engineering ingenuity, and their effectiveness is evaluated through extensive simulations, the focus of these schemes is on algorithmic design, without knowing the fundamental computational complexity of the localization process.

The localization problem is similar to the graph embedding problem. In [24], Saxe shows that testing the embeddability of weighted graphs in k-space is strongly **NP**-hard. However, the graphs he considers are general graphs. In sensor network localization, since only nodes who are within a communication range can measure their relative distances, the graphs formed by connecting each pair of nodes who can measure each other's distance are better modeled as unit disk graphs [7]. Such constraints could have the potential of allowing computationally efficient localization algorithms to be designed. For example, in [25], Biswas and Ye show that network localization in unit disk graphs can be formulated as a semidefinite programming problem and thus can be efficiently solved. A condition of their algorithm, however, is that the graphs are densely connected. More specifically, their algorithm requires that $\Omega(n^2)$ pairs of nodes know their relative distances, where n is the number of sensor nodes in the network. However, for a general network, it is enough for the localization process to have a unique solution when $O(n)$ pairs of nodes know their distances, if certain conditions are satisfied [9].

Researchers are still looking for efficient algorithms that work for sparse networks. Such algorithms are of great importance, because in the limit as a network with bounded communication range and fixed sensor density grows, the number of known distance pairs grows only linearly in the number of nodes.

In this paper, we show that localization in sparse networks is **NP**-hard. The main result and its consequences are presented in Section 2. The proof of the main result, which makes up the bulk of the paper, is presented in Section 3. Finally, a discussion of conclusions and open problems is presented in Section 4.

2   Localization

The **localization problem** considered in this paper is to reconstruct the positions of a set of sensors given the distances between any pair of sensors that are within some unit disk radius r of each other. Some of the sensors may be **beacons**, sensors with known positions, but our impossibility results are not affected much by whether beacons are available. To avoid precision issues involving irrational distances, we assume that the input to the problem is presented with the distances squared. If we make the further assumption that all sensors have integer coordinates, all distances will be integers as well.

For the main result, we consider a decision version of the localization problem, which we call **UNIT DISK GRAPH RECONSTRUCTION** . This problem essentially asks if a particular graph with given edge lengths can be physically realized as a unit disk graph with a given disk radius in two dimensions.

The input is a graph G where each edge uv of G is labeled with an integer ℓ_{uv}^2, the square of its length, together with an integer r^2 that is the square of the radius of a unit disk. The output is "yes" or "no" depending on whether there exists a set of points in R^2 such that the distance between u and v is ℓ_{uv} whenever uv is an edge in G and exceeds r whenever uv is not an edge in G.

Our main result, is that UNIT DISK GRAPH RECONSTRUCTION is **NP**-hard, based on a reduction from CIRCUIT SATISFIABILITY. The constructed graph for a circuit with m wires has $O(m^2)$ vertices and $O(m^2)$ edges, and the number of solutions to the resulting localization problem is equal to the number of satisfying assignments for the circuit. In each solution to the localization problem, the points can be placed at integer coordinates, and the entire graph fits in an $O(m)$-by-$O(m)$ rectangle, where the constants hidden by the asymptotic notation are small. The construction also permits a constant fraction of the nodes to be placed at known locations.

Formally, we show:

Theorem 1. *There is a polynomial-time reduction from CIRCUIT SATISFIABILITY to UNIT DISK GRAPH RECONSTRUCTION, in which there is a one-to-one correspondence between satisfying assignments to the circuit and solutions to the resulting localization problem.*

The proof of Theorem 1 is given in Section 3. A consequence of this result is:

Corollary 1. *There is no efficient algorithm that solves the localization problem for sparse sensor networks in the worst case unless $P = NP$.*

Proof. Suppose that we have a polynomial-time algorithm that takes as input the distances between sensors from an actual placement in R^2, and recovers the original position of the sensors (relative to each other, or to an appropriate set of beacons). Such an algorithm can be used to solve UNIT DISK GRAPH RECONSTRUCTION by applying it to an instance of the problem (that may or may not have a solution). After reaching its polynomial time bound, the algorithm will either have returned a solution or not. In the first case, we can check if the solution returned is consistent with the distance constraints in the UNIT DISK GRAPH RECONSTRUCTION instance in polynomial time, and accept if and only if the check succeeds. In the second case, we can reject the instance. In both cases we have returned the correct answer for UNIT DISK GRAPH RECONSTRUCTION. □

It might appear that this result depends on the possibility of ambiguous reconstructions, where the position of some points is not fully determined by the known distances. However, if we allow randomized reconstruction algorithms, a similar result holds even for graphs that have unique reconstructions

Corollary 2. *There is no efficient randomized algorithm that solves the localization problem for sparse sensor networks that have unique reconstructions unless $RP = NP$.*

Proof. The proof of this claim is by use of the well-known construction of Valiant and Vazirani [26], which gives a randomized Turing reduction from 3SAT to UNIQUE SATISFIABILITY. The essential idea of this reduction is that randomly fixing some of the inputs to the 3SAT problem reduces the number of potential solutions, and repeating the process eventually produces a 3SAT instance with a unique solution with high probability.   □

Finally, because the graph constructed in the proof of Theorem 1 uses only points with integer coordinates, even an approximate solution that positions each point to within a distance $\epsilon < 1/2$ of its correct location can be used to find the exact locations of all points by rounding each coordinate to the nearest integer. Since the construction uses a fixed value for the unit disk radius r (the natural scale factor for the problem), we have

Corollary 3. *The results of Corollary 1 and Corollary 2 continue to hold even for algorithms that return an approximate location for each point, provided the approximate location is within $\epsilon \cdot r$ of the correct location, where ϵ is a fixed constant.*

What we do *not* know at present is whether these results continue to hold for solutions that have large positional errors but that give edge lengths close to those in the input. Our suspicion is that edge-length errors accumulate at most polynomially across the graph, but we have not yet carried out the error analysis necessary to prove this. If our suspicion is correct, we would have:

Conjecture 1. The results of Corollary 1 and Corollary 2 continue to hold even for algorithms that return an approximate location for each point, provided the relative error in edge length for each edge is bounded by ϵ/n^c for some fixed constant c.

3   Proof of Theorem 1

The proof is by constructing a family of graphs that represent arbitrary Boolean circuits, where the relative positions of certain rigid bodies within the graph correspond to signals and NOT and AND gates are built out of additional constraints between these bodies. This gives a reduction to UNIT DISK GRAPH RECONSTRUCTION from CIRCUIT SATISFIABILITY, which shows that determining whether a solution to a particular localization problem in this class exists is **NP**-hard.

The construction requires the development of several preliminary tools. Our main tool will be a two-state hinge structure that permits adjacent parts of the graph to be shifted relative to each other by $\pm h$ for some length h. These hinges will then be used to hook together a two-dimensional grid of rigid bodies whose rows and columns act like wires, transmitting one bit each across the grid. Additional applications of the hinge give junctions between rows and columns (forcing a row and column to carry the same bit, similar to a junction in a circuit), negation, and AND gates (via an implication mechanism that enforces $x \rightarrow y$ for chosen variables x and y).

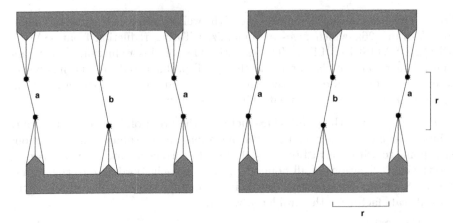

Fig. 1. The basic binary hinge. The top and bottom structures, including the prongs supporting the hinges, are rigid; points within the shaded polygons and the edges between them are not depicted explicitly, but are assume to be sufficiently dense to force a unique placement of the hinge endpoints. The three edges between the top and bottom endpoints have lengths a, $b > a$, and a, and enforce that the vertical separation between the two bodies is constant and that the horizontal offset is $\pm h$ for some h. The disk radius r is shown for scale.

A binary hinge. We begin by describing a **binary hinge**, which is a connection between two rigid bodies that enforces a fixed offset between them in the direction of the hinge and an offset of $\pm h$ for some h in the perpendicular direction. As depicted in Figure 1, the hinge consists of three edges between fixed points on the surfaces of the bodies, with the inner edge longer than the outer two. Lemma 1 below describes the exact structure of the hinge and its properties.

Lemma 1. *Fix points A at coordinates $(0,0)$, B at coordinates $(m,0)$, and K at coordinates $(m/2, -k)$ for some $k > 0$. Let $h < \ell\sqrt{3}$ and let C and D be points at distance $m > 1$ from each other such that C is at distance $\sqrt{\ell^2 + h^2}$ from A, D is at distance $\sqrt{\ell^2 + h^2}$ from B, and their midpoint M is at distance $\sqrt{(\ell + k)^2 + h^2}$ from K, where $\ell < m$. Then either C is at coordinates $(-h, \ell)$ and D is at coordinates $(m - h, \ell)$, or C is at $(+h, \ell)$ and D is at $(m + h, \ell)$.*

Proof. It is easily verified that $C = (\pm h, \ell)$ is at distance $\sqrt{\ell^2 + h^2}$ from $A = (0,0)$, that $D = (m \pm h, \ell)$ is at distance $\sqrt{\ell^2 + h^2}$ from $B = (m,0)$, and that $M = (m/2 \pm h, \ell)$ is at distance $\sqrt{(\ell + k)^2 + h^2}$ from $K = (m/2, -k)$. It remains to show that only these two solutions exist.

First we consider only the constraints of the outer two edges. The distance constraints imply that C and D lie on circles of radius $\sqrt{\ell^2 + h^2}$ centered at A and B respectively. Without loss of generality, let $\sqrt{\ell^2 + h^2} = 1$ and consider Figure 2. For a fixed position of D, C lies on a radius-m circle centered at D; this circle intersects the radius-1 circle centered at A in exactly two points C and C'.

In the first case, $ABDC$ is a parallelogram and the midpoint M of C and D lies on a radius-1 circle centered at the midpoint of A and B. This will give rise to our two solutions.

The second case is more complicated. We will first show that the midpoint M' is close to the AB line. Here we have C' at distance m from D, but $ABDC'$ is not a parallelogram but is instead a quadrilateral that crosses itself. To see that the AB and DC' line segments do in fact cross, observe that $\triangle AC'D$ and $\triangle DBA$ both have edges of length 1, m, and x where x is the length of AD and are thus congruent. Within $\triangle AC'D$, the angle $\angle DAC'$ is opposite a longer side than the angle $\angle C'DA$ and is thus larger; it follows that $\angle DAC'$ is larger than $\angle BAD$ and thus C' and D lie on opposite sides of the AB line. The y-coordinate of M' is thus the average of two quantities y_1 and y_2 which lie between -1 and +1 with opposite sign, and so it lies in the closed interval $[-1/2, 1/2]$.

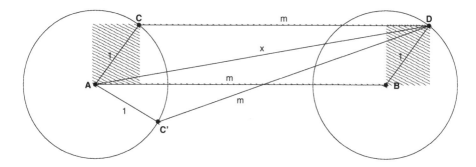

Fig. 2. Two configurations of the outer edges of the binary hinge. In the figure, the edges are scaled to have length 1. In the first configuration, the two edges AC and BD are parallel, producing the parallelogram $ACDB$. In the second, the left edge drops to the bottom part of the circle to produce the self-crossing quadrilateral $AC'DB$. The point C' always lies below the AB line provided $m > 1$, which implies that the midpoint of the $C'D$ edge has a y-coordinate no greater than $1/2$.

We have shown that any midpoint M of C and D lies either on the circle or within the shaded region depicted in Figure 3. The middle edge of the hinge requires that it also lie on the radius $\sqrt{(\ell + k)^2 + h^2}$ circle with center K. This circle intersects the previous circle in precisely the points $(m/2 \pm h, \ell)$ described in the lemma; because $h < \ell\sqrt{3}$, we have $\sqrt{\ell^2 + h^2} < \sqrt{4\ell^2} = 2\ell$, and so ℓ is at least $1/2$ under our scaling assumption. In particular the K-centered circle intersects inner circle above $1/2$ and thus avoids the shaded region, yielding only the two solutions $(m/2 \pm h, \ell)$. □

The circuit grid. We now use the binary hinge construction to construct a grid of horizontal and vertical wires, represented by rows and columns of rigid bodies in a two-dimensional array. The core of the construction is depicted in Figure 4: each body (which consists of the hinge endpoints and enough internal nodes

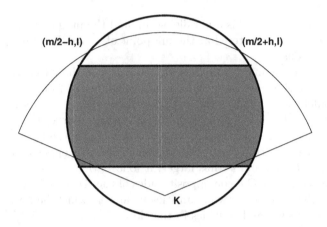

Fig. 3. Applying the middle hinge constraint. The middle hinge forces the midpoint M to lie on the circle centered at K (here depicted as a wedge). The proof of Lemma 1 shows that M must either lie on the depicted circle or within the shaded region; if the intersection points of the two circles lie outside the shaded region they are the only solutions.

and internal bracing to assure rigidity) is connected by binary hinges to its four immediate neighbors in the grid. This enforces a fixed vertical spacing ℓ between the centers of the body and its neighbors above and below, and the same fixed horizontal spacing ℓ between the centers of the body and its neighbors to the left and the right. These fixed spacings mean that the horizontal offsets propagate across rows and vertical offsets propagate across columns. At the same time, the hinges allow different bodies in the same row or column to have different vertical or horizontal offsets, respectively.

To organize the grid into wires, we anchor it to an L-shaped external frame as depicted in Figure 5. This frame fixes the horizontal and vertical offsets of every other row and column to some fixed value that we take to be 0. The remaining rows and columns are free to shift by $\pm h$ relative to these fixed rows and columns where h is the offset parameter of the hinges. These rows act as wires transmitting binary signals; we think of a column as having the value 1 when shifted up and 0 when shifted down, and of a row as having the value 1 when shifted right and 0 when shifted left.

The offset h can be chosen somewhat arbitrarily, but should be substantially less than half the disk radius (and thus the length of the hinges) to allow for double-offset hinges to be used later without exceeding the $\ell\sqrt{3}$ limit in Lemma 1. To minimize the size of the graph, we will assume that the offset h, the disk radius r, and the separation between centers of grid elements ℓ are all small constants. Reasonable values that permit the construction to work without its components colliding are $h = 1$, $r = 10$, and $\ell = 150$; with these values, all of the nodes in the graph can be placed at integer coordinates and no extraneous edges creep in. In

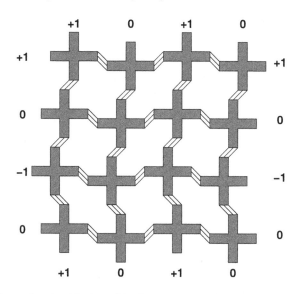

Fig. 4. Two-dimensional grid of rigid bodies connected by binary hinges. Because each hinge fixes the separation between the bodies it connects but allows sideways offsets in either direction, row and column offsets propagate from one side of the grid to the other without interference.

the figures, both the offset and separation of each hinge is greatly exaggerated for visibility.

The fixed rows and columns divide the blocks into four categories, depending on whether the horizontal and/or vertical positions are fixed. Those for which both positions are fixed act as a fixed extension of the frame that will be used later to create circuit connections between horizontal and vertical wires. Those for which neither position is fixed act as crossing points (with no connection) between the wires.

The circuit grid by itself provides only wires but no connections between them. To build a working circuit, we need to build (a) junctions between horizontal and vertical wires, allowing arbitrary routing of signals; (b) negation; and (c) some sort of two-input gate (we will build AND). We start by describing how to build a junction between wires.

Junctions. For a junction, we want to enforce that a horizontal wire is in its rightward position (representing a 1 bit) if and only if the joined vertical wire is in its upward position (also representing a 1 bit). We do so by placing a hinge with offset $\pm h\sqrt{2}$ and separation $\ell\sqrt{2}$ at a 45° angle between the crossing body for the two wires and an adjacent fixed body, as shown in Figure 6. Note that these irrational lengths are a result of the 45° rotation; the nodes that make up the hinge are still at integer coordinates.

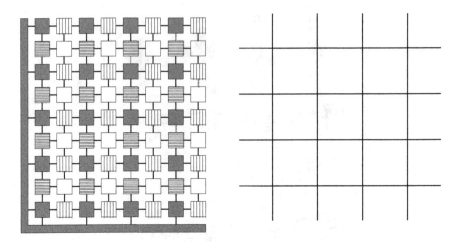

Fig. 5. Two-dimensional grid of rigid bodies (squares) joined by binary hinges (dark lines) and with every other row and column anchored by binary hinges to an outer frame (L-shaped body) is used to simulate a circuit consisting of horizontal and vertical wires with no junctions (right-hand figure). The horizontally cross-hatched bodies are offset by $\pm h$ horizontally and correspond to horizontal wires. The vertically cross-hatched bodies are offset by $\pm h$ vertically and correspond to vertical wires. Bodies without cross-hatching are wire crossings that can move both horizontally and vertically; shaded bodies are fixed.

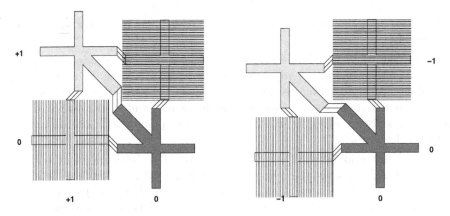

Fig. 6. Junction between two wires. The hinge between the fixed body (at lower right) and the crossing point (at upper left) forces the crossing point into either its $(+1, +1)$ position or its $(-1, -1)$ position, which guarantees that the horizontal and vertical wires carry the same signal.

NOT gates. To force two adjacent horizontal wires to have opposite values, we place a double-width hinge with offset $\pm 2h$ between them at the right edge of the grid as shown in Figure 7. This permits one to be offset by $+h$ and the other by $-h$, but prevents both from being offset by the same amount. We will assume hereafter that the horizontal wires are paired off in this way with the two members of each pair representing some variable and its negation. To extract these variables for use we create a junction that sends a copy up one of the vertical wires.

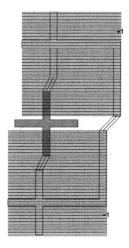

Fig. 7. Negation. Double-width hinge between top and bottom rows allows only $+h$ and $-h$ or $-h$ and $+h$ as offsets.

Implications. At the top of the grid, we place implications. These are structures built out of single-width hinges (with offset $\pm h$) that enforce $x \to y$ by permitting only the offset combinations $(-1, -1)$, $(-1, +1)$, and $(+1, +1)$. This restriction is enforced by setting the x side of the hinge h units higher than the right side; the sides of the hinge are then at vertical offsets $(0, -h)$, $(0, h)$, and $(2h, h)$ in the three permissible configurations. The configuration where x is one and y is zero would have offsets $(2h, -h)$, which is not permitted by the hinge. By placing junctions appropriately we can create an implication between any two variables represented by horizontal wires. See Figure 8.

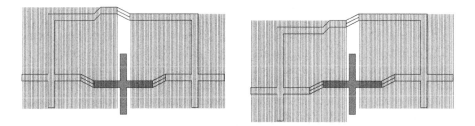

Fig. 8. Implication. If left-hand column is up, right-hand column is also forced up. If left-hand column is down, right-hand column may be up or down.

Implication is a rather weak primitive, but it is enough to build AND gates.

AND gates. An AND gate is a structure that forces some variable z to equal $x \wedge y$. We can build an AND gate out of four implications: $z \to x$, $z \to y$, $\overline{x} \to \overline{z}$, and $\overline{y} \to \overline{z}$, which between them allow precisely those settings of x, y, and z for which $z = x \wedge y$.

Since we have both AND gates and negation, we can build OR gates as well using DeMorgan's rule $x \vee y = \overline{\neg x \wedge \neg y}$. This is enough to build arbitrary Boolean circuits. In particular, we have a:

Reduction from CIRCUIT SATISFIABILITY. Given a Boolean circuit C, construct a graph G using the mechanisms described previously, where each wire (including input and output wires) and its negation is represented by a row of rigid bodies, and each AND gate is implemented using eight columns paired together with four implication structures. Any solution to the localization problem for the distances in this graph corresponds to an assignment of bits to the wires in C that is consistent with the behavior of the gates in C; to enforce that the output of the circuit is 1, anchor the appropriate row to the L-shaped frame to force it into the 1 position. Now there is a solution to the localization problem if and only if there is a satisfying assignment for the circuit.

We thus have a polynomial-time reduction from CIRCUIT SATISFIABILITY to localization, and Theorem 1 is proved.

4   Conclusions and Open Problems

We have shown that the localization problem is hard in the worst case for sparse graphs unless $\mathbf{P} = \mathbf{NP}$ or $\mathbf{RP} = \mathbf{NP}$, if certain mild forms of approximation are permitted. This worst-case result for sparse graphs stands in contrast to results that show that localization is possible for dense graphs [25] or with high probability for random geometric graphs [9]. The open questions that remain are where the boundary lies between our negative result and these positive results. In particular:

- Is there an efficient algorithm for *approximate* localization in sparse graphs, either by permitting moderate errors on distances or by permitting the algorithm to misplace some small fraction of the sensors?
- Given that the difficulty of the problem appears to be strongly affected by the density of nodes (and the resulting number of known distance pairs), what minimum density is necessary to allow localization in the worst case?
- How are these results affected by more natural assumptions about communications ranges, allowing different maximum distances between adjacent nodes or the possibility of placing small numbers of high-range beacons?

Answers to any of these questions would be an important step toward producing practical localization algorithms.

References

1. Joe Albowicz, Alvin Chen, and Lixia Zhang. Recursive position estimation in sensor networks. In *Proceedings of the 9th International Conference on Network Protocols '01*, pages 35–41, Riverside, CA, November 2001.
2. Paramvir Bahl and Venkata N. Padmanabhan. RADAR: An in-building RF-based user location and tracking system. In *Proceedings of IEEE INFOCOM '00*, pages 775–784, Tel Aviv, Israel, March 2000.
3. Pratik Biswas and Yinyu Ye. Semide
   nite programming for ad hoc wireless sensor network localization. In Feng Zhao and Leonidas Guibas, editors, *Proceedings of Third International Workshop on Information Processing in Sensor Networks*, Berkeley, CA, April 2004.
4. N. Bulusu, J. Heidemann, and D. Estrin. GPS-less low-cost outdoor localization for very small devices. *IEEE Personal Communications Magazine*, 7(5):28–34, October 2000.
5. Srdan Capkun, Maher Hamdi, and Jean-Pierre Hubaux. GPS-free positioning in mobile ad-hoc networks. In *HICSS*, 2001.
6. Krishna Chintalapudi, Ramesh Govindan, Gaurav Sukhatme, and Amit Dhariwal. Ad-hoc localization using ranging and sectoring. In *Proceedings of IEEE INFOCOM '04*, Hong Kong, China, April 2004.
7. B. N. Clark, C. J. Colbourn, and D. S. Johnson. Unit disk graphs. *Discrete Mathematics*, 86:165–177, 1990.
8. L. Doherty, K. S. J. Pister, and L. E. Ghaoui. Convex position estimation in wireless sensor networks. In *Proceedings of IEEE INFOCOM '01*, pages 1655–1633, Anchorage, AK, April 2001.
9. T. Eren, D. Goldenberg, W. Whitley, Y.R. Yang, S. Morse, B.D.O. Anderson, and P.N. Belhumeur. Rigidity, computation, and randomization of network localization. In *Proceedings of IEEE INFOCOM '04*, Hong Kong, China, April 2004.
10. Deborah Estrin, Ramesh Govindan, John S. Heidemann, and Satish Kumar. Next century challenges: Scalable coordination in sensor networks. In *Proceedings of The Fifth International Conference on Mobile Computing and Networking (Mobicom)*, pages 263–270, Seattle, WA, November 1999.
11. G. H. Forman and J. Zahorjan. The challenges of mobile computing. *IEEE Computer*, 27(4):38–47, Apr. 1994.
12. L. Girod and D. Estrin. Robust range estimation using acoustic and multimodal sensing. In *IEEE/RSI Int. Conf. on Intelligent Robots and Systems (IROS)*, 2001.
13. T. He, C. Huang, B. Blum, J. Stankovic, and T. Abdelzaher. Range-free localization schemes in large scale sensor networks. In *Proceedings of The Ninth International Conference on Mobile Computing and Networking (Mobicom)*, pages 81–95, San Diego, CA, Sep 2003.
14. Jeffrey Hightower and Gaetano Borriella. Location systems for ubiquitous computing. *IEEE Computer*, 34(8):57–66, 2001.
15. B. Hofmann-Wellenhof, H. Lichtenegger, and J. Collins. *Global Positioning System: Theory and Practice, Fourth Edition*. Springer-Verlag, 1997.
16. Xiang Ji. Sensor positioning in wireless ad-hoc sensor networks with multidimensional scaling. In *Proceedings of IEEE INFOCOM '04*, Hong Kong, China, April 2004.
17. Brad Karp and H. T. Kung. GPSR: Greedy perimeter stateless routing for wireless networks. In *Proceedings of The Sixth International Conference on Mobile Computing and Networking (Mobicom)*, Boston, MA, August 2000.

18. Seapahn Meguerdichian, Farinaz Koushanfar, Gang Qu, and Miodrag Potkonjak. Exposure in wireless ad-hoc sensor networks. In *Proceedings of The Seventh International Conference on Mobile Computing and Networking (Mobicom)*, Rome, Italy, July 2001.

19. D. Niculescu and B. Nath. Ad-hoc positioning system. In *Proceedings of IEEE Globecom 2001*, November 2001.

20. Dragos Niculescu and Badri Nath. Ad hoc positioning system (APS) using AOA. In *Proceedings of IEEE INFOCOM '03*, San Francisco, CA, April 2003.

21. Nissanka B. Priyantha, Anit Chakraborty, and Hari Balakrishnan. The cricket location-support system. In *Proceedings of The Sixth International Conference on Mobile Computing and Networking (Mobicom)*, pages 32–43, Boston, MA, August 2000.

22. C. Savarese, J. Rabay, and K. Langendoen. Robust positioning algorithms for distributed ad-hoc wireless sensor networks. In *USENIX Technical Annual Conference*, Monterey, CA, June 2002.

23. Andreas Savvides, Chih-Chieh Han, and Mani B. Strivastava. Dynamic fine-grained localization in ad-hoc networks of sensors. *In Proceedings of The Seventh International Conference on Mobile Computing and Networking (Mobicom)*, pages 166–179, Rome, Italy, July 2001.

24. J.B. Saxe. Embeddability of weighted graphs in k-space is strongly NP-hard. In *Proceedings of the 17th Allerton Conference in Communications, Control and Computing*, pages 480–489, 1979.

25. Yi Shang and Wheeler Ruml. Improved MDS-based localization. In *Proceedings of IEEE INFOCOM '04*, Hong Kong, China, April 2004.

26. Leslie G. Valiant and Vijay V. Vazirani. NP is as easy as detecting unique solutions. *Theoretical Computer Science*, 47(1):85–93, 1986.

27. Roy Want, Andy Hopper, Veronica Falcão, and Jonathan Gibbons. The active badge location system. Technical Report 92.1, Olivetti Research Ltd. (ORL), 24a Trumpington Street, Cambridge CB2 1QA, 1992.

28. A. Ward, A. Jones, and A. Hopper. A new location technique for the active office. *IEEE Personal Communications*, 4(5):42–47, 1997.

29. M. Weiser. Some computer science problems in ubiquitous computing. *Communications of ACM*, July 1993.

30. J. Werb and C. Lanzl. Designing a positioning system for finding things and people indoors. *IEEE Spectrum*, 35(9):71–78, October 1998.

A Distributed TDMA Slot Assignment Algorithm for Wireless Sensor Networks

Ted Herman[1] and Sébastien Tixeuil[2]

[1] University of Iowa, `ted-herman@uiowa.edu`
[2] LRI – CNRS UMR 8623 & INRIA Grand Large, Université Paris-Sud XI, France,
`tixeuil@lri.fr`

Abstract. Wireless sensor networks benefit from communication protocols that reduce power requirements by avoiding frame collision. Time Division Media Access methods schedule transmission in slots to avoid collision, however these methods often lack scalability when implemented in *ad hoc* networks subject to node failures and dynamic topology. This paper reports a distributed algorithm for TDMA slot assignment that is self-stabilizing to transient faults and dynamic topology change. The expected local convergence time is $O(1)$ for any size network satisfying a constant bound on the size of a node neighborhood.

1 Introduction

Collision management and avoidance are fundamental issues in wireless network protocols. Networks now being imagined for sensors [24] and small devices [3] require energy conservation, scalability, tolerance to transient faults, and adaptivity to topology change. Time Division Media Access (TDMA) is a reasonable technique for managing wireless media access, however the priorities of scalability and fault tolerance are not emphasized by most previous research. Recent analysis [8] of radio transmission characteristics typical of sensor networks shows that TDMA may not substantially improve bandwidth when compared to randomized collision avoidance protocols, however fairness and energy conservation considerations remain important motivations. In applications with predictable communication patterns, a sensor may even power off the radio receiver during TDMA slots where no messages are expected; such timed approaches to power management are typical of the sensor regime.

Emerging models of *ad hoc* sensor networks are more constrained than general models of distributed systems, especially with respect to computational and communication resources. These constraints tend to favor simple algorithms that use limited memory. A few constraints of some sensor networks can be helpful: sensors may have access to geographic coordinates and a time base (such as GPS provides), and the density of sensors in an area can have a known, fixed upper bound. The question we ask in this paper is how systems can distributively obtain

S. Nikoletseas and J. Rolim (Eds.): ALGOSENSORS 2004, LNCS 3121, pp. 45–58, 2004.

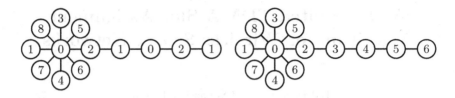

Fig. 1. Two solutions to distance-two coloring

a TDMA assignment of slots to nodes, given the assumptions of synchronized clocks and a bounded density (where density is interpreted to be a fixed upper bound on the number of immediate neighbors in the communication range of any node). In practice, such a limit on the number of neighbors in range of a node has been achieved by dynamically attenuating transmission power on radios. Our answers to the question of distributively obtaining a TDMA schedule are partial: our results are not necessarily optimum, and although the algorithms we present are self-stabilizing, they are not optimally designed for all cases of minor disruptions or changes to a stabilized sensor network.

Before presenting our results, it may be helpful for the reader to consider the relation between TDMA scheduling and standard problems of graph coloring (since these topics often found in textbooks on network algorithms for spatial multiplexing). Algorithmic research on TDMA relates the problem of timeslot assignment to minimal graph coloring where the coloring constraint is typically that of ensuring that no two nodes within distance two have the same color (the constraint of distance two has a motivation akin to the well known hidden terminal problem in wireless networks). This simple reduction of TDMA times-lot assignment neglects some opportunities for time division: even a solution to minimum coloring does not necessarily give the best result for TDMA slot assignment. Consider the two colorings shown in Figure 1, which are minimum distance-two colorings of the same network. We can count, for each node p, the size of the set of colors used within its distance-two neighborhood (where this set includes p's color); this is illustrated in Figure 2 for the respective colorings of Figure 1. We see that some of the nodes find more colors in their distance-two neighborhoods in the second coloring of Figure 1. The method of slot allocation in Section 6 allocates larger bandwidth share when the number of colors in distance-two neighborhoods is smaller. Intuitively, if some node p sees $k < \lambda$ colors in its distance-two neighborhood, then it should have at least a $1/(k+1)$ share of bandwidth, which is superior to assigning a $1/(\lambda + 1)$ share to each color. Thus the problem of optimum TDMA slot assignment is, in some sense, harder than optimizing the global number of colors.

Contributions. The main issues for our research are dynamic network configurations, transient fault tolerance and scalability of TDMA slot assignment algorithms. Our approach to both dynamic network change and transient fault events is to use the paradigm of self-stabilization, which ensures the system

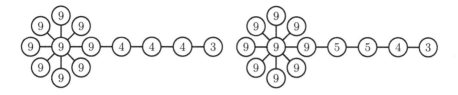

Fig. 2. Number of colors used within distance two

state converges to a valid TDMA assignment after any transient fault or topology change event. Our approach to scalability is to propose a randomized slot assignment algorithm with $O(1)$ expected *local* convergence time. The basis for our algorithm is, in essence, a probabilistically fast clustering technique (which could be exploited for other problems of sensor networks). The expected time for *all* nodes to have a valid TDMA assignment is not $O(1)$; our view is that stabilization over the entire network is an unreasonable metric for sensor network applications; we discuss this further in the paper's conclusion. Our approach guarantees that after stabilization, if nodes crash, TDMA collision may occur only locally (in the distance-three neighborhood of the faults).

Related Work. The idea of self-stabilizing TDMA has been developed in [13, 14] for model that is more restricted than ours (a grid topology where each node knows its location). Algorithms for allocating TDMA time slots and FDMA frequencies are formulated as vertex coloring problems in a graph [17]. Let the set of vertex colors be the integers from the range $0..\lambda$. For FDMA the colors (f_v, f_w) of neighboring vertices (v, w) should satisfy $|f_v - f_w| > 1$ to avoid interference. The standard notation for this constraint is $L(\ell_1, \ell_2)$: for any pair of vertices at distance $i \in \{1, 2\}$, the colors differ by at least ℓ_i. The coloring problem for TDMA is: let $L'(\ell_1, \ell_2)$ be the constraint that for any pair of vertices at distance $i \in \{1, 2\}$, the colors differ by at least $\ell_i \bmod (\lambda + 1)$. (This constraint represents the fact that time slots wrap around, unlike frequencies.) The coloring constraint for TDMA is $L'(1, 1)$. Coloring problems with constraints $L(1, 0)$, $L(0, 1)$, $L(1, 1)$, and $L(2, 1)$ have been well-studied not only for general graphs but for many special types of graphs [2,11,18]; many such problems are NP-complete and although approximation algorithms have been proposed, such algorithms are typically not distributed. (The related problem finding a minimum dominating set has been shown to have a distributed approximation using constant time [12], though it is unclear if the techniques apply to self-stabilizing coloring.) Self-stabilizing algorithms for $L(1, 0)$ have been studied in [5,21,19,20, 7], and for $L(1, 1)$ in [6]. Our algorithms borrow from techniques of self-stabilizing coloring and renaming [6,7], which use techniques well-known in the literature of parallel algorithms on PRAM models [15]. To the extent that the sensor network model is synchronous, some of these techniques can be adapted; however working out details when messages collide, and the initial state is unknown, is not an entirely trivial task. This paper is novel in the sense that it composes self-stabilizing algorithms for renaming and coloring for a base model that has only probabilistically correct communication, due to the possibility of collisions at the media access layer. Also, our coloring uses a constant number of colors for the $L(1,1)$ problem, while the previous self-stabilizing solution to this problem uses n^2 colors. Due to space constraints, proofs are delegated to [9].

2   Wireless Network, Program Notation

The system is comprised of a set V of nodes in an *ad hoc* wireless network, and each node has a unique identifier. Communication between nodes uses a low-power radio. Each node p can communicate with a subset $N_p \subseteq V$ of nodes determined by the range of the radio signal; N_p is called the neighborhood of node p. In the wireless model, transmission is omnidirectional: each message sent by p is effectively broadcast to all nodes in N_p. We also assume that communication capability is bidirectional: $q \in N_p$ iff $p \in N_q$. Define $N_p^1 = N_p$ and for $i > 1$, $N_p^i = N_p^{i-1} \cup \{r \mid (\exists q : q \in N_p^{i-1} : r \in N_q)\}$ (call N_p^i the distance-i neighborhood of p). Distribution of nodes is sparse: there is some known constant δ such that for any node p, $|N_p| \leq \delta$. (Sensor networks can control density by powering off nodes in areas that are too dense, which is one aim of topology control algorithms.)

Each node has fine-grained, real-time clock hardware, and all node clocks are synchronized to a common, global time. Each node uses the same radio frequency (one frequency is shared spatially by all nodes in the network) and media access is managed by CSMA/CA: if node p has a message ready to transmit, but is receiving some signal, then p does not begin transmission until it detects the absence of signal; and before p transmits a message, it waits for some random period (as implemented, for instance, in [23]). We assume that the implementation of CSMA/CA satisfies the following: there exists a constant $\tau > 0$ such that the probability of a frame transmission without collision is at least τ (this corresponds to typical assumptions for multiaccess channels [1]; the independence of τ for different frame transmissions indicates our assumption of an underlying memoryless probability distribution in a Markov model).

Notation. We describe algorithms using the notation of guarded assignment statements: $G \to S$ represents a guarded assignment, where G is a predicate of the local variables of a node, and S is an assignment to local variables of the node. If predicate G (called the *guard*) holds, then assignment S is executed, otherwise S is skipped. Some guards can be event predicates that hold upon the event of receiving a message: we assume that all such guarded assignments execute atomically when a message is received. At any system state where a given guard G holds, we say that G is *enabled* at that state. The $[]$ operator is the nondeterministic composition of guarded assignments; $([]q : q \in M_p : G_q \to S_q)$ is a closed-form expression of $G_{q1} \to S_{q1} [] G_{q2} \to S_{q2} [] \cdots [] G_{qk} \to S_{qk}$, where $M_p = \{q_1, q_2, \ldots, q_k\}$.

Execution Semantics. The life of computing at every node consists of the infinite repetition of finding a guard and executing its corresponding assignment or skipping the assignment if the guard is *false*. Generally, we suppose that when a node executes its program, all statements with *true* guards are executed in some constant time (done, for example, in round-robin order).

Shared Variable Propagation. A certain subset of the variables at any node are designated as *shared* variables. Nodes periodically transmit the values of their shared variables, based on a timed discipline. Given the discipline of repeated transmission of shared variables, each node can have a cached copy of the value of a shared variable for any neighbor. This cached copy is updated atomically upon receipt of a message carrying a new value for the shared variable.

Model Construction. Our goal is to provide an implementation of a general purpose, collision-free communication service. This service can be regarded as a transformation of the given model of Section 2 into a model without collisions. This service simplifies application programming and can reduce energy require-ments for communication (messages do not need to be resent due to collisions). Let $\mathcal{T}$ denote the task of transforming the model of Section 2 into a collision-free model.

To solve $\mathcal{T}$ it suffices to assign each node a color and use node colors as the schedule for a TDMA approach to collision-free communication [17]. Even before colors are assigned, we use a schedule that partitions radio time into two parts: one part is for TDMA scheduling of application messages and the other part is reserved for the messages of the algorithm that assigns colors and time slots to nodes. The following diagram illustrates such a schedule, in which each TDMA part has five slots. Each overhead part is, in fact, a fixed-length slot in the TDMA schedule.

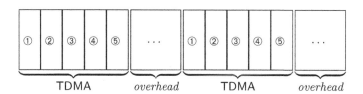

The programming model, including the technique for sharing variables, refers to message and computation activity in the overhead parts. Whereas CSMA/CA is used to manage collisions in the overhead slots, the remaining TDMA slots do not use random delay. During initialization or after a dynamic topology change, frames may collide in the TDMA slots, but after the slot assignment algorithm self-stabilizes, collisions do not occur in the TDMA slots.

With respect to any given node v, a solution $\mathcal{T}$ is *locally stabilizing* with con-vergence time t if, for any initial system state, after at most t time units, every subsequent system state satisfies the property that any transmission by v during its assigned slot(s) is free from collision. Solution $\mathcal{T}$ is *globally stabilizing* with convergence time t if, for every initial state, after at most t time units, every subsequent system state has the property that all transmissions during assigned slots are free from collision. For randomized algorithms, these definitions are modified to specify expected convergence times (all stabilizing randomized algorithms we consider are probabilistically convergent in the Las Vegas sense). When the qualification (local or global) is omitted, convergence times for local stabilization are intended for the presented algorithms.

Several primitive services that are not part of the initial model can simplify the design and expression of $\mathcal{T}$'s implementation. All of these services need to be self-stabilizing. Briefly put, our plan is to develop a sequence of algorithms that enable TDMA implementation. These algorithms are: neighborhood-unique naming, maximal independent set, minimal coloring, and the assignment of time slots from colors. In addition, we rely on neighborhood services that update cached copies of shared variables.

Neighborhood Identification. We do not assume that a node p has built-in knowledge of its neighborhood N_p or its distance-three neighborhood N_p^3. This is because the type of network under considering is *ad hoc*, and the topology dynamic. Therefore some algorithm is needed so that a node can refer to its neighbors. We describe first how a node p can learn of N_p^2, since the technique can be extended to learn N_p^3 in a straightforward way.

Each node p can represent N_p^i for $i \in 1..3$ by a list of identifiers learned from messages received at p. However, because we do not make assumptions about the initial state of any node, such list representations can initially have arbitrary data. Let L be a data type for a list of up to δ items of the form $a : A$, where a is an identifier and A is a set of up to δ identifiers. Let sL_p be a shared variable of type L. Let message type mN with field of type L be the form of messages transmitted for sL_p. Let L_p be a private variable of a type that is an augmentation of L – it associates a real number with each item: $age(a : A)$ is a positive real value attached to the item.

Function $\mathsf{update}(L_p, a : A)$ changes L_p to have new item information: if L_p already has some item whose first component is a, it is removed and replaced with $a : A$ (which then has age zero); if L_p has fewer than δ items and no item with a as first component, then $a : A$ is added to L_p; if L_p has already δ items and no item with a as first component, then $a : A$ replaces some item with maximal age.

Let maxAge be some constant designed to be an upper limit on the possible age of items in L_p. Function $\mathsf{neighbors}(L_p)$ returns the set

$$\{\, q \mid q \neq p \,\wedge\, (\exists\, (a : A) : (a : A) \in L_p : a = q)\, \}$$

Given these variable definitions and functions, we present the algorithm for neighborhood identification.

> N0: *receive* $mN(a : A)$ $\rightarrow$ $\mathsf{update}(L_p, a : A \setminus \{p\})$

N1: $([] (a : A) \in L_p : age(a : A) > \mathsf{maxAge} \rightarrow L_p := L_p \setminus (a : A))$
N2: $true \rightarrow sL_p := (p : \mathsf{neighbors}(L_p))$

The constant maxAge should be tuned to safely remove old or invalid neighbor data, yet to retain current neighbor information by receiving new mN messages before age expiration. This is an implementation issue beyond of the scope of this paper: our abstraction of the behavior of the communication layer is the assumption that, eventually for any node, the guard of N1 remains *false* for any $(a : A) \in L_p$ for which $a \in N_p$. By a similar argument, eventually each node p correctly has knowledge of N_p^2 and N_p^3 as well as N_p. In all subsequent sections, we use N_p^i for $i \in 1..3$ as constants in programs with the understanding that such neighborhood identification is actually obtained by the stabilizing protocol described above.

Building upon L_p, cached values of the shared variables of nodes in N_p^i, for $i \in 1..3$, can be maintained at p; erroneous cache values not associated with any node can be discarded by the aging technique. We use the following notation in the rest of the paper: for node p and some shared variable var_q of node $q \in N_p^3$, let $\boxtimes var_q$ refer to the cached copy of var_q at p.

Problem Definition. Let $\mathcal{T}$ denote the task of assigning TDMA slots so that each node has some assigned slot(s) for transmission, and this transmission is guaranteed to be collision-free. We seek a solution to $\mathcal{T}$ that is distributed and self-stabilizing in the sense that, after some transient failure or reconfiguration, node states may not be consistent with the requirements of collision-free communication and collisions can occur; eventually the algorithm corrects node states to result in collision-free communication.

3   Neighborhood Unique Naming

An algorithm providing neighborhood-unique naming gives each node a name distinct from any of its N^3-neighbors. This may seem odd considering that we already assume that nodes have unique identifiers, but when we try to use the identifiers for certain applications such as coloring, the potentially large namespace of identifiers can cause scalability problems. Therefore it can be useful to give nodes smaller names, from a constant space of names, in a way that ensures names are locally unique.

The problem of neighborhood unique naming can be considered as an N^3-coloring algorithm and quickly suggests a solution to $\mathcal{T}$. Since neighborhood unique naming provides a solution to the problem of $L(1,1)$ coloring, it provides a schedule for TDMA. This solution would be especially wasteful if the space of unique identifiers is larger than $|V|$. It turns out that having unique identifiers within a neighborhood can be exploited by other algorithms to obtain a minimal N^2-coloring, so we present a simple randomized algorithm for N^3-naming.

Our neighborhood unique naming algorithm is roughly based on the randomized technique described in [6], and introduces some new features. Define $\Delta = \lceil \delta^t \rceil$ for some $t > 3$; the choice of t to fix constant Δ has two competing motivations discussed at the end of this section. We call Δ the *namespace*. Let shared variable Id_p have domain $0..\Delta$; variable Id_p is the *name* of node p. Another variable is used to collect the names of neighboring nodes: $Cids_p = \{\boxtimes Id_q \mid q \in N_p^3 \setminus \{p\}\}$. Let $\mathsf{random}(S)$ choose with uniform probability some element of set S. Node p uses the following function to compute Id_p:

$$\mathsf{newId}(Id_p) = \begin{cases} Id_p & \text{if } Id_p \notin Cids_p \\ \mathsf{random}(\Delta \setminus Cids_p) & \text{otherwise} \end{cases}$$

The algorithm for unique naming is the following.

$\mathsf{N3}:\ true\ \rightarrow\ Id_p := \mathsf{newId}(Id_p)$

Define $\mathsf{Uniq}(p)$ to be the predicate that holds iff (i) no name mentioned in $Cids_p$ is equal to Id_p, (ii) for each $q \in N_p^3$, $q \neq p$, $Id_q \neq Id_p$, (iii) for each $q \in N_p^3$, one name in $Cids_q$ equals Id_p, (iv) for each $q \in N_p^3$, $q \neq p$, the equality $\boxtimes Id_p = Id_p$ holds at node q, and (v) no cache update message *en route* to p conveys a name that would update $Cids_p$ to have a name equal to Id_p. Predicate $\mathsf{Uniq}(p)$ states that p's name is known to all nodes in N_p^3 and does not conflict with any name of a node q within N_q^3, nor is there a cached name liable to update $Cids_p$ that conflicts with p's name. A key property of the algorithm is the following: $\mathsf{Uniq}(p)$ is a stable property of the execution. This is because after $\mathsf{Uniq}(p)$ holds, any node q in N_p^3 will not assign Id_q to equal p's name, because N3 avoids names listed in the cache of distance-three neighborhood names – this stability property is not present in the randomized algorithm [6]. The property $(\forall r :\ r \in R :\ \mathsf{Uniq}(r))$ is similarly stable for any subset R of nodes. In words, once a name becomes established as unique for all the neighborhoods it belongs to, it is stable. Therefore we can reason about a Markov model of executions by showing that the probability of a sequence of steps moving, from one stable set of ids to a larger stable set, is positive.

Lemma 1. *Starting from any state, there is a constant, positive probability that* $\mathsf{Uniq}(p)$ *holds within constant time.*

Corollary 1. *The algorithm self-stabilizes with probability 1 and has constant expected local convergence time.*

Using the names assigned by N3 is a solution to $L(1,1)$ coloring, however using Δ colors is not the basis for an efficient TDMA schedule. The naming obtained by the algorithm does have a useful property. Let P be a path of t distinct nodes, that is, $P = p_1, p_2, \ldots, p_t$. Define predicate $Up(P)$ to hold if $id_{p_i} < id_{p_j}$ for each $i < j$. In words, $Up(P)$ holds if the names along the path P increase.

Lemma 2. *Every path P satisfying $Up(P)$ has fewer than $\Delta + 1$ nodes.*

This lemma shows that the simple coloring algorithm gives us a property that node identifiers do not have: the path length of any increasing sequence of names is bounded by a constant. Henceforth, we suppose that node identifiers have this property, that is, we treat N_p^i as if the node identifiers are drawn from the namespace[1] of size Δ.

4   Leaders via Maximal Independent Set

Simple distance two coloring algorithms may use a number of colors that is wastefully large. Our objective is to find an algorithm that uses a reasonable number of colors and completes, with high probability, in constant time. We observe that an assignment to satisfy distance two coloring can be done in constant time given a set of neighborhood leader nodes distributed in the network. The leaders dictate coloring for nearby nodes. The coloring enabled by this method is minimal (not minimum, which is an NP-hard problem). An algorithm selecting a maximal independent set is our basis for selecting the leader nodes.

Let each node p have a boolean shared variable ℓ_p. In an initial state, the value of ℓ_p is arbitrary. A legitimate state for the algorithm satisfies $(\forall p : \ p \in V : \ \mathcal{L}_p)$, where

$$\mathcal{L}_p \equiv (\ell_p \Rightarrow (\forall q : \ q \in N_p : \ \neg \ell_q)) \wedge (\neg \ell_p \Rightarrow (\exists q : \ q \in N_p : \ \ell_q))$$

Thus the algorithm should elect one leader (identified by the ℓ-variable) for each neighborhood. As in previous sections, $\boxtimes \ell_p$ denotes the cached copy of the shared variable ℓ_p.

R1: $(\forall q : \ q \in N_p : \ q > p) \ \rightarrow \ \ell_p := \textit{true}$
R2: $(\square q : \ q \in N_p : \ \boxtimes \ell_q \wedge q < p \ \rightarrow \ \ell_p := \textit{false})$
R3: $(\exists q : \ q \in N_p : \ q < p) \wedge (\forall q : \ q \in N_p \wedge (q > p \vee \neg \boxtimes \ell_q)) \ \rightarrow \ \ell_p := \textit{true}$

Although the algorithm does not use randomization, its convergence technically remains probabilistic because our underlying model of communication uses CSMA/CA based on random delay. The algorithm's progress is therefore guaranteed with probability 1 rather than by deterministic means.

Lemma 3. *With probability 1 the algorithm R1-R3 converges to a solution of maximal independent set; the convergence time is $O(1)$ if each timed variable propagation completes in $O(1)$ time.*

---

[1] There are two competing motivations for tuning the parameter t in $\Delta = \delta^t$. First, t should be large enough to ensure that the choice made by newId is unique with high probability. In the worst case, $|N_p^3| = \delta^3 + 1$, and each node's cache can contain δ^3 names, so a choosing $t \approx 6$ could be satisfactory. Generally, larger values for t decrease the expected convergence time of the neighborhood unique naming algorithm. On the other hand, smaller values of t will reduce the constant Δ, which will reduce the convergence time for algorithms described in subsequent sections.

5   Leader Assigned Coloring

Our method of distance-two coloring is simple: colors are assigned by the leader nodes given by maximal independent set output. The following variables are introduced for each node p:

$color_p$ is a number representing the color for node p.
$min\ell_p$ is meaningful only for p such that $\neg\ell_p$ holds: it is intended to satisfy

$$min\ell_p = \min \{ q \mid q \in N_p \wedge \boxtimes \ell_q \}$$

In words, $min\ell_p$ is the smallest id of any neighbor that is a leader. Due to the uniqueness of names in N_p^3, the value $min\ell_p$ stabilizes to a unique node.

$spectrum_p$ is a set of pairs (c,r) where c is a color and r is an id. Pertaining only to nonleader nodes, $spectrum_p$ should contain $(color_p, min\ell_p)$ and $(\boxtimes color_q, \boxtimes min\ell_q)$ for each $q \in N_p$.

$setcol_p$ is meaningful only for p such that ℓ_p holds. It is an array of colors indexed by identifier: $setcol_p[q]$ is p's preferred color for $q \in N_p$. We consider $color_p$ and $setcol_p[p]$ to be synonyms for the same variable. In the algorithm we use the notation $setcol_p[A] := B$ to denote the parallel assignment of a set of colors B based on a set of indices A. To make this assignment deterministic, we suppose that A can be represented by a sorted list for purposes of the assignment; B is similarly structured as a list.

dom_p for leader p is computed to be the nodes to which p can give a preferred color; these include any $q \in N_p$ such that $min\ell_q = p$. We say for $q \in dom_p$ that p *dominates* q.

f is a function used by each leader p to compute a set of unused colors to assign to the nodes in dom_p. The set of *used* colors for p is

$$\{ c \mid (\exists q, r : q \in N_p \wedge (c,r) \in spectrum_q \wedge r < p) \}$$

That is, used colors with respect to p are those colors in N_p^2 that are already assigned by leaders with smaller identifiers than p. The complement of the *used* set is the range of possible colors that p may prefer for nodes it dominates. Let f be the function to minimize the number of colors preferred for the nodes of dom_p, ensuring that the colors for dom_p are distinct, and assigning smaller color indices (as close to 0 as possible) preferentially. Function f returns a list of colors to match the deterministic list of dom_p in the assignment of R5.

R4: $\ell_p \rightarrow dom_p := \{p\} \cup \{q \mid q \in N_p \wedge \boxtimes min\ell_q = p \}$
R5: $\ell_p \rightarrow setcol_p[dom_p] := f(\{c \mid \exists q : q \in N_p \wedge r < p \wedge (c,r) \in \boxtimes spectrum_q \})$
R6: $true \rightarrow min\ell_p := \min \{ q \mid q \in N_p \cup \{p\} \wedge \boxtimes \ell_q \}$
R7: $\neg\ell_p \rightarrow color_p := \boxtimes setcol_r[p]$, where $r = min\ell_p$
R8: $\neg\ell_p \rightarrow spectrum_p := (color_p, min\ell_p) \cup \bigcup \{ (c,r) \mid$
      $(\exists q, c, r : q \in N_p : c = \boxtimes color_q \wedge r = \boxtimes min\ell_q) \}$

Lemma 4. *The algorithm R4-R8 converges to a distance-two coloring, with probability 1; the convergence time is $O(1)$ if each timed variable propagation completes in $O(1)$ time.*

Due to space restrictions, we omit the proof that the resulting coloring is minimal (which follows from the construction of f to be locally minimum, and the essentially sequential assignment of colors along paths of increasing names).

6 Assigning Time Slots from Colors

Given a distance-two coloring of the network nodes, the next task is to derive time slot assignments for each node for TDMA scheduling. Our starting assumption is that each node has equal priority for assigning time slots, *ie*, we are using an unweighted model in allocating bandwidth. Before presenting an algorithm, we have two motivating observations.

First, the algorithms that provide coloring are local in the sense that the actual number of colors assigned is not available in any global variable. Therefore to assign time slots consistently to all nodes apparently requires some additional computation. In the first solution of Figure 1, both leftmost and rightmost nodes have color 1, however only at the leftmost node is it clear that color 1 should be allocated one ninth of the time slots. Local information available at the rightmost node might imply that color 1 should have one third of the allocated slots.

The second observation is that each node p should have about as much bandwidth as any other node in N_p^2. This follows from our assumption that all nodes have equal priority. Consider the N_p^2 sizes shown in Figure 2 that correspond to the colorings of 1. The rightmost node p in the first coloring has three colors in its two-neighborhood, but has a neighbor q with four colors in its two-neighborhood. It follows that q shares bandwidth with four nodes: q's share of the bandwidth is at most $1/4$, whereas p's share is at most $1/3$. It does not violate fairness to allow p to use $1/3$ of the slot allocation if these slots would otherwise be wasted. Our algorithm therefore allocates slots in order from most constrained (least bandwidth share) to least constrained, so that extra slots can be used where available.

To describe the algorithm that allocates media access time for node p, we introduce these shared variables and local functions.

$base_p$ stabilizes to the number of colors in N_p^2. The value $base_p^{-1} = 1/base_p$ is used as a constraint on the share of bandwidth required by p in the TDMA slot assignment.

$itvl_p$ is a set of intervals of the form $[x, y)$ where $0 \leq x < y \leq 1$. For allocation, each unit of time is divided into intervals and $itvl_p$ is the set of intervals that node p can use to transmit messages. The expression $|[x, y)|$ denotes the time-length of an interval.

$g(b, S)$ is a function to assign intervals, where S is a set of subintervals of $[0, 1)$. Function $g(b, S)$ returns a maximal set T of subintervals of $[0, 1)$ that are disjoint and also disjoint from any element of S such that $(\sum_{a \in T} |a|) \leq b$.

To simplify the presentation, we introduce S_p as a private (nonshared) variable.

R9: $true \rightarrow base_p := | \{ \boxtimes color_q \mid q \in N_p^2 \} |$
R10: $true \rightarrow S_p := \bigcup \{ \boxtimes itvl_q \mid q \in N_p^2 \; \wedge$
       $(\boxtimes base_q > base_p \; \vee$
       $(\boxtimes base_q = base_p \; \wedge \; \boxtimes color_q < color_p)) \}$
R11: $true \rightarrow itvl_p := g(base_p^{-1}, S_p)$

Lemma 5. *With probability 1 the algorithm R9–R11 converges to an allocation of time intervals such that no two nodes within distance two have conflicting time intervals, and the interval lengths for each node p sum to $|\{ color_q \mid q \in N_p^2 \}|^{-1}$; the expected convergence time of R9-R11 is $O(1)$ starting from any state with stable and valid coloring.*

It can be verified of R9-R11 that, at a fixed point, no node $q \in N_p^2$ is assigned a time that overlaps with interval(s) assigned to p; also, all available time is assigned (there are no leftover intervals). A remaining practical issue is the conversion from intervals to a time slot schedule: a discrete TDMA slot schedule will approximate the intervals calculated by g.

7 Assembly

Given the component algorithms of Sections 2–6, the concluding statement of our result follows.

Theorem 1. *The composition of N0–N3 and R1–R11 is a probabilistically self-stabilizing solution to $\mathcal{T}$ with $O(1)$ expected local convergence time.*

8 Conclusion

Sensor networks differ in characteristics and in typical applications from other large scale networks such as the Internet. Sensor networks of extreme scale (hundreds of thousands to millions of nodes) have been imagined [10], motivating scalability concerns for such networks. The current generation of sensor networks emphasizes the *sensing* aspect of the nodes, so services that aggregate data and report data have been emphasized. Future generations of sensor networks will have significant actuation capabilities. In the context of large scale sensor/actuator networks, end-to-end services can be less important than regional and local services. Therefore we emphasize local stabilization time rather than global stabilization time in this paper, as the local stabilization time is likely to be more important for scalability of TDMA than global stabilization time. Nonetheless, the question of global stabilization time is neglected in previous sections. We speculate that global stabilization time will be sublinear in the diameter of the network (which could be a different type of argument for scalability of our constructions, considering that end-to-end latency would be linear in the network diameter even after stabilization). Some justification for our speculation is the following: if the expected local time for convergence is $O(1)$ and underlying probability assumptions are derived from Bernoulli (random name selection) and Poisson (wireless CSMA/CA) distributions, then these distributions can be approximately bounded by exponential distributions with constant means. Exponential distributions define half-lives for populations of convergent processes (given asymptotically large populations), which is to say that within some constant time γ, the expected population of processes that have not converged is halved; it would follow that global convergence is $O(\lg n)$.

We close by mentioning two important open problems. Because sensor networks can be deployed in an *ad hoc* manner, new sensor nodes can be dynamically thrown into a network, and mobility is also possible, the TDMA algorithm we propose could have a serious disadvantage: introduction of just one new node could disrupt the TDMA schedules of a sizable part of a network before the system stabilizes. Even if the stabilization time is expected to be $O(1)$, it may be that better algorithms could contain the effects of small topology changes with less impact than our proposed construction. One can exploit normal notifications of topology change as suggested in [4], for example. Another interesting question is whether the assumption of globally synchronized clocks (often casually defended by citing GPS availability in literature of wireless networks) is really needed for self-stabilizing TDMA construction; we have no proof at present that global synchronization is necessary.

References

1. D. Bertsekas, R. Gallager. *Data Networks*, Prentice-Hall, 1987.
2. H. L. Bodlaender, T. Kloks, R. B. Tan, J. van Leeuwen. Approximations for λ-coloring of graphs. University of Utrecht, Department of Computer Science, Technical Report 2000-25, 2000 (25 pages).
3. D. E. Culler, J. Hill, P. Buonadonna, R. Szewczyk, A. Woo. A network-centric approach to embedded software for tiny devices. In *Proceedings of Embedded Software, First International Workshop EMSOFT 2001*, Springer LNCS 2211, pp. 114-130, 2001.
4. S Dolev and T Herman. Superstabilizing protocols for dynamic distributed systems. *Chicago Journal of Theoretical Computer Science*, 3(4), 1997.
5. S. Ghosh, M. H. Karaata. A self-stabilizing algorithm for coloring planar graphs. Distributed Computing 7:55-59, 1993.

6. M. Gradinariu, C. Johnen. Self-stabilizing neighborhood unique naming under unfair scheduler. In *Euro-Par'01 Parallel Processing, Proceedings,* Springer LNCS 2150, 2001, pp. 458-465.

7. M. Gradinariu, S. Tixeuil. Self-stabilizing vertex coloration of arbitrary graphs. In *4th International Conference On Principles Of DIstributed Systems, OPODIS'2000,* 2000, pp. 55-70.

8. M. Haenggi, X. Liu. Fundamental throughput limits in Rayleigh fading sensor networks. *in submission,* 2003.

9. T Herman, S Tixeuil A distributed TDMA slot assignment algorithm for wireless sensor networks. Technical Report, University of Iowa Department of Computer Science, 2004. CoRR Archive Number cs.DC/0405042.

10. J. Kahn, R. Katz, and K. Pister. Next century challenges: mobile networking for "smart dust". In *Proceedings of the Fifth Annual International Conference on Mobile Computing and Networking (MOBICOM '99),* 1999.

11. S. O. Krumke, M. V. Marathe, S. S. Ravi. Models and approximation algorithms for channel assignment in radio networks. *Wireless Networks* 7(6 2001):575-584.

12. F. Kuhn, R. Wattenhofer. Constant-time distributed dominating set approximation. In *Proceedings of the Twenty-Second ACM Symposium on Principles of Distributed Computing,* (PODC 2003), pp. 25-32, 2003.

13. S. S. Kulkarni, U. Arumugam. Collision-free communication in sensor networks. In *Proceedings of Self-Stabilizing Systems, 6th International Symposium,* Springer LNCS 2704, 2003, pp. 17-31.

14. S. S. Kulkarni and U. Arumugam. Transformations for Write-All-With-Collision Model. In *Proceedings of the 7th International Conference on Principles of Distributed Systems (OPODIS),* Springer LNCS, 12/03. (Martinique, French West Indies, France).

15. M. Luby. A simple parallel algorithm for the maximal independent set problem. *SIAM Journal on Computing* 15:1036-1053, 1986.

16. M. Mizuno, M. Nesterenko. A transformation of self-stabilizing serial model programs for asynchronous parallel computing environments. *Information Processing Letters* 66 (6 1998):285-290.

17. S. Ramanathan. A unified framework and algorithm for channel assignment in wireless networks. *Wireless Networks* 5(2 1999):81-94.

18. S. Ramanathan, E. L. Lloyd. Scheduling algorithms for multi-hop radio networks. *IEEE/ACM Transactions on Networking,* 1(2 1993):166-177.

19. S. Shukla, D. Rosenkrantz, and S. Ravi. Developing self-stabilizing coloring algorithms via systematic randomization. In *Proceedings of the International Workshop on Parallel Processing,* pages 668–673, Bangalore, India, 1994. Tata-McGrawhill, New Delhi.

20. S. Shukla, D. Rosenkrantz, and S. Ravi. Observations on self-stabilizing graph algorithms for anonymous networks. In *Proceedings of the Second Workshop on Self-stabilizing Systems (WSS'95),* pages 7.1–7.15, 1995.

21. S. Sur and P. K. Srimani. A self-stabilizing algorithm for coloring bipartite graphs. *Information Sciences,* 69:219–227, 1993.

22. G. Tel. *Introduction to Distributed Algorithms,* Cambridge University Press, 1994.

23. A. Woo, D. Culler. A transmission control scheme for media access in sensor networks. In *Proceedings of the Seventh International Conference on Mobile Computing and Networking (Mobicom 2001),* pp. 221-235, 2001.

24. F. Zhao, L. Guibas (Editors). *Proceedings of Information Processing in Sensor Networks, Second International Workshop, IPSN 2003,* Springer LNCS 2634. April, 2003.

Balanced Data Gathering in Energy-Constrained Sensor Networks*

Emil Falck[1], Patrik Floréen[2], Petteri Kaski[1],
Jukka Kohonen[2], and Pekka Orponen[1]

[1] Laboratory for Theoretical Computer Science, P.O. Box 5400, FI-02015 Helsinki
University of Technology, Finland. `firstname.lastname@hut.fi`.

[2] Helsinki Institute for Information Technology/Basic Research Unit, Department of
Computer Science, P.O. Box 26, FI-00014 University of Helsinki, Finland.
`firstname.lastname@cs.helsinki.fi`.

Abstract. We consider the problem of gathering data from a wireless
multi-hop network of energy-constrained sensor nodes to a common base
station. Specifically, we aim to balance the total amount of data received from the sensor network during its lifetime against a requirement
of sufficient coverage for all the sensor locations surveyed. Our main
contribution lies in formulating this balanced data gathering task and
in studying the effects of balancing. We give an LP network flow formulation and present experimental results on optimal data routing designs
also with impenetrable obstacles between the nodes. We then proceed to
consider the effect of augmenting the basic sensor network with a small
number of auxiliary relay nodes with less stringent energy constraints.
We present an algorithm for finding approximately optimal placements
for the relay nodes, given a system of basic sensor locations, and compare
it with a straightforward grid arrangement of the relays.

1 Introduction

Wireless networks consisting of a large number of miniature electromechanical devices with sensor, computing and communication capabilities are rapidly
becoming a reality, due to the accumulation of advances in digital electronics, wireless communications and microelectromechanical technology [1,14,22].
Prospective applications of such devices cover a wide range of domains [1,7,9,
10].

One generic type of application for sensor networks is the continuous monitoring of an extended geographic area at relatively low data rates [1,5]. The information provided from all points of the sensor field is then gathered via multi-hop
communications to a base station for further processing. We are here envisaging
a scenario where environmental data are frequently and asynchronously collected

* Research supported by the Academy of Finland, Grants 202203 (J. Kohonen, P.
  Floréen), 202205 (E. Falck, P. Kaski) and 204156 (P. Orponen); and by the Foundation of Technology (Tekniikan Edistämissäätiö), Helsinki, Finland (P. Kaski).

S. Nikoletseas and J. Rolim (Eds.): ALGOSENSORS 2004, LNCS 3121, pp. 59–70, 2004.

over an area, and all information is to be gathered for later postprocessing of best possible quality, including detection of possibly faulty data. This means that data aggregation [17,18] cannot be employed.

Significant design constraints are imposed by multi-hop routing and the limited capabilities and battery power available at the sensor nodes. A number of recent papers have addressed e.g. optimal sensor placement [6,11,12,16,21] and energy-efficient routing designs and protocols [8,13,15,17,19,20,24] with the objective of lifetime maximization [3,4,17,21].

We envisage the sensor placement to be fixed beforehand, either by an application expert according to the needs of the particular application at hand, or randomly, for example by scattering them from an airplane. For the sake of achieving a comprehensive view of the whole area to be monitored, not only should the total amount of data received at the base station be maximized, but the different sensors should be able to get through to the base station some minimum amount of data. The main contribution of this paper lies in formulating and studying this balanced data gathering problem.

The idea of incorporating a certain balancing requirement on the data gathering has also recently been proposed in [19,20] and in [11]. In [19,20] the authors put forth a more general model of information extraction that takes into account the nonlinear relation between transmission power and information rate. Our problem formulation can be seen as a linearized, computationally feasible version of this approach. Another difference between [19,20] and our work pertains to the expression of the balancing, or fairness, requirement. Article [11] considers the problem of maximizing the lifetime of a sensor network, and explicates this task in terms of an integer program that counts the number of "rounds" the network is operational, assuming that each sensor sends one data packet in each round. This formulation entails a strict fairness condition among the sensors, requiring them all to send exactly equal amounts of data. We allow an adjustable trade-off between maximizing the total amount of data received at the base station and the minimum amount of data received from each sensor. Moreover our program formulation does not require integer variables.

In Sect. 2 we formulate the problem of balanced data gathering as a linear program (LP). In Sect. 3 we present some experiments on routing designs and data flows resulting from various balancing requirements. Although we do not address the issue of ideal placement of the sensor nodes, in Sect. 4 we do consider the effects, from the point of view of our balanced data gathering measure, of augmenting a given sensor network by a small number of auxiliary radio relay nodes with higher battery power levels. The locations of these relays may be chosen at will, and we present and compare two heuristics for determining good relay locations to optimize the network behavior.

It needs to be noted that our linear program based solution relies on information about all the energy costs of transmitting and receiving a unit of data, and the data rates and energy supplies of all nodes. However, knowledge of node locations as such is not required, and our model readily adjusts to obstacles and other deviations from simple radio-link models as long as the energy costs of transmission can be determined by the nodes themselves, either by simply probing at different power levels, or using more sophisticated means such as a Received Signal Strength Indicator (RSSI) [23].

Our linear program formulation also requires all the information to be available at a single location. This assumption is realistic only if the operation time is long and the amount of control traffic small. Otherwise, the protocol must be able to decide on the basis of local information. Our results thus provide an upper bound on what is actually achievable using distributed protocols, with local and imprecise information.

2   Optimization of Balanced Data Gathering

We consider a network consisting of n sensor nodes, m relay nodes and a base station, all with predetermined locations, except for the relay nodes whose locations may be changed. For simplicity, we index the nodes so that the base station has index 1. The set of all nodes is denoted as $V = B \cup S \cup R$, where B, S, and R denote the sets consisting of the base station node, sensors, and relays, respectively.

Each node $i \in V$ has an initial energy supply of e_i units; as a special case, we set $e_1 = \infty$. The mission of the network is to gather data generated at the sensor (source) nodes to the base station (sink) node under these energy constraints, during the desired operation time T.

We assume that the sensors generate data asynchronously and in such small unit packets that the process can be modeled by assigning to each of the sensor nodes an "offered data rate" parameter s_i, $i \in S$. The energy cost of forwarding a unit of data from node i to node j is given by a parameter d_{ij} and the cost of receiving a unit of data is given by a parameter c. We also assume the transmission rates to be low enough, so that collisions and signal interference can be ignored in the model.

Our model places no restrictions on the values of the parameters d_{ij} and c. As an example, in the commonly used simple radio-link models [23], d_{ij} would be taken to be proportional to $c_t + D_{ij}^{\alpha}$, where c_t corresponds to the energy consumed by the transmitter electronics and D_{ij}^{α} corresponds to the energy consumed by the transmit amplifier to achieve an acceptable signal-to-noise ratio at the receiving node; D_{ij} is the physical distance between nodes i and j and the exponent α, $2 \leq \alpha \lesssim 4$, models the decay of the radio signal in the ambient medium. The cost c corresponds to the energy consumed by the receiver electronics.

Thus, if we introduce flow variables f_{ij} indicating the rate of data forwarded from each node i to node j, the energy constraints in the network can be expressed as $(\sum_j d_{ij} f_{ij} + \sum_j c f_{ji}) \cdot T \leq e_i$, for all $i \in V$.

Because of the energy constraints in the network, the sensors cannot usually productively achieve their full offered data rates; thus we introduce variables r_i indicating the actual "achieved data rate" at each sensor $i \in S$. One goal of the data flow design for the network is to maximize the total, or equivalently, the average achieved data rate $(1/n)\sum_{i\in S} r_i$. However, taking this as the singular objective may lead to the "starvation" of some of the sensor nodes: typically, the average data rate objective is maximized by data flows that only forward data generated close to the sink, and do not allocate any energy towards relaying data generated at distant parts of the network.

To counterbalance this tendency, we add a "minimum achieved rate" variable ℓ, with the constraints $r_i \geq \ell$ for all $i \in S$, and try to maximize this simultaneously with the average data rate. The trade-off between these two conflicting objectives is determined by a parameter λ, $0 \leq \lambda \leq 1$, where value $\lambda = 0$ gives all weight to the average achieved rate objective, and value $\lambda = 1$ to the minimum achieved rate objective. The combined objective F_λ, see (1), will subsequently be called the "balanced data rate," or "balanced rate."

Different sensors may submit different types of data. At each unit of time, the average amount of data transmitted from one sensor might be one bit, and from another sensor ten bits; however, the one bit may be equally valuable for the application as the other sensor's ten bits. As a generalization, we can assign weights w_i to the data rates from different sensors according to their importance. A natural choice is $w_i = 1/s_i$, which normalizes the data rates of all sensors to the interval $[0, 1]$, and expresses the idea that an equal proportion of each sensor's offered data should be transmitted. For simplicity, we have used equal offered data rates and equal weights in our experiments.

Our model can now be formulated as the following linear program, which can then be solved using standard techniques. Note that a linear programming approach is taken also in [11,17,21,24].

$$\max \quad F_\lambda := (1 - \lambda)\frac{1}{n}\sum_{i\in S} w_i r_i + \lambda\ell \tag{1}$$

$$\text{s.t.} \quad \sum_{j\in V} f_{1j} = 0,$$

$$\sum_{j\in V} f_{ij} = r_i + \sum_{j\in V} f_{ji}, \qquad i \in S,$$

$$\sum_{j\in V} f_{ij} = \sum_{j\in V} f_{ji}, \qquad i \in R,$$

$$\sum_{j\in V} Td_{ij}f_{ij} + \sum_{j\in V} Tcf_{ji} \leq e_i, \quad i \in V,$$

$$r_i \leq s_i, \qquad i \in S,$$

$$w_i r_i \geq \ell, \qquad i \in S,$$

$$f_{ij} \geq 0, \qquad i, j \in V,$$

$$f_{ii} = 0, \qquad i \in V.$$

A flow matrix f_{ij}, obtained as a solution to this linear program, can easily be used to route approximately r_i unit-size data packets from each source node $i \in S$ to the sink node 1, assuming that all the r_i and f_{ij} values are large. At each node k, simply forward the first $\lfloor f_{k1} \rfloor$ packets to node 1 (the sink), the next $\lfloor f_{k2} \rfloor$ packets to node 2, the next $\lfloor f_{k3} \rfloor$ packets to node 3, and so on. A somewhat more elegant solution is to randomize the routing strategy, so that each incoming packet at node i is forwarded to node j with probability $f_{ij} / \sum_k f_{ik}$.

3   Experimental Results

As already mentioned, we do not address the problem of optimal sensor placement, but take the sensor locations as given. Since our focus is on studying the effect of the balancing factor λ on the resulting data flows and sensor data rates in the network, we choose in most of our experiments to place the sensor nodes in a regular grid. This eliminates the coincidental effects arising in, e.g., a random node placement from the variations in the distances between the nodes. However, we illustrate also the case of random node placement in Sect. 3.2.

3.1   Node Placement in a Regular Grid

In our first experiments, we place 100 sensors in a 10×10 grid in a square of dimensions 1 km $\times$ 1 km and the base station at the middle of one of the sides of the square. All sensors have an energy constraint of 20 J and an offered amount of data of 100 Mbits during the operation of the network. If the time of operation is $T = 10^6$ seconds, this translates to an average offered data rate of 100 bit/s. Transmission and reception costs are computed as $d_{ij} = 100$ nJ/bit $+$ 0.01 nJ/bit/m$^2 \cdot D_{ij}^2$ and $c = 100$ nJ/bit. These values are comparable to those in [4,15].

The effects of the balancing factor λ on the resulting flows and rates in the network are illustrated in Figs. 1, 2 and 3. In these examples, the offered rates are relatively high, making the network heavily energy-constrained: the achieved data rates are limited more by the network's ability to transport data than by the sensors' ability to generate it.

If no balancing is required ($\lambda = 0$), the objective is simply to maximize the average data rate from all sensors. Since the sensors nearest to the base station can provide a large contribution to the average by sending at relatively low energy cost, the optimum solution indeed allocates most of the network's resources to this goal. Very little data is received from the most distant sensors. As higher balancing factors are used ($\lambda = 0.5$, $\lambda = 1$), the distant nodes get a bigger share of the network's transport resources – and accordingly, the area is more evenly covered by observations. This comes at a cost of reducing the average rate.

Depending on the characteristics of the network, the effect of balancing can be quite large. With no balancing the achieved rates are over 12 bit/s on the average, but below 2 bit/s for the most distant nodes. With full balancing, all sensors get to transmit at an equal rate of about 7.4 bit/s.

An alternative approach would be to limit the data rates of the nearby nodes from above (e.g. as in [19]), in order to prevent them from sending unnecessarily

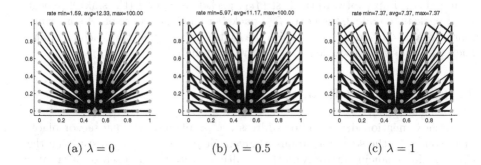

(a) $\lambda = 0$ (b) $\lambda = 0.5$ (c) $\lambda = 1$

Fig. 1. Sensor network with 100 sensor nodes placed in a grid: Optimal flow solutions for different values of the balancing factor λ. Line widths are proportional to the square root of the flow.

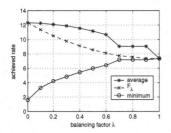

Fig. 2. Average, balanced and minimum data rates (bit/s), as a function of the balancing factor.

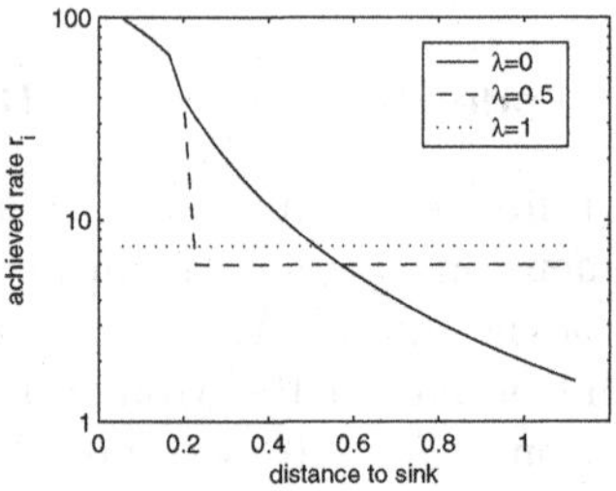

Fig. 3. Achieved data rates r_i for each individual sensor (logarithmic scale).

large amounts of data, and to save energy for the data from the more distant nodes. However, our approach of using a lower limit more directly models the intuitive goal of gathering enough data equally from all parts of the area to be monitored.

3.2 Random Node Placement

To illustrate random node placement, we made a second set of experiments; see Fig. 4. Network parameters are otherwise the same as before, but the 100 nodes were placed randomly in the square area, using uniform distribution. The effects of balancing are very similar to those described above for the regular grid.

3.3 Addition of Obstacles

Our LP formulation readily allows additional constraints to be included. For instance, we could limit the transmission powers of the nodes or the channel capacities of the links.

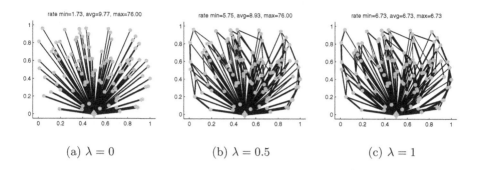

(a) $\lambda = 0$ (b) $\lambda = 0.5$ (c) $\lambda = 1$

Fig. 4. Sensor network with 100 randomly placed sensor nodes: Optimal flow solutions for different values of the balancing factor λ.

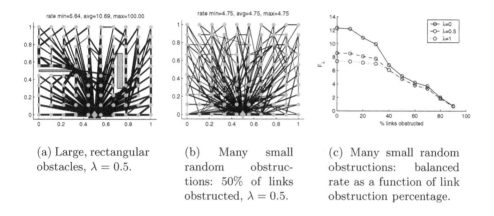

(a) Large, rectangular obstacles, $\lambda = 0.5$.

(b) Many small random obstructions: 50% of links obstructed, $\lambda = 0.5$.

(c) Many small random obstructions: balanced rate as a function of link obstruction percentage.

Fig. 5. Experiments with networks with nonuniform signal propagation.

Since our model allows arbitrary energy costs D_{ij}, it is not restricted to idealized transmission conditions like the popular unit disk model. We can study situations where there are large impenetrable obstacles within the area, as illustrated in Fig. 5(a). This is done by assigning an infinite energy cost to any link that intersects an obstacle.

In Figs. 5(b) and 5(c) we illustrate the effects of a different kind of nonuniform propagation. Instead of large obstacles, a random subset of all links is made unavailable by assigning them an infinite cost. This corresponds to small obstructions scattered throughout the area.

As expected, the achieved balanced rate F_λ decreases as more links are obstructed. However, the performance of the network degrades gracefully. As seen in the curve for $\lambda = 1$, even with half of all links randomly obstructed, the simulated network is able to transport a minimum rate of 4.7 bit/s from every sensor node, or about 65 % of the minimum rate of 7.4 in the unobstructed network.

4   The Effect of Relay Nodes

The performance of the sensor network can be improved by augmenting the network by a number of auxiliary relay nodes. Unlike sensor nodes, whose locations are assumed to be predetermined, the locations of the relay nodes may be chosen to optimize the network performance. The relay nodes do not generate data themselves, but are solely used for forwarding data to other nodes in the network. Furthermore, the relay nodes may have considerably higher initial energy supplies than the sensor nodes.

In this section we consider the effect of relay nodes on network performance, and present and compare two simple techniques for determining good relay node locations.

4.1   Relay Node Placement Methods: Grid and Incremental Optimization

In the case of a square area, a straightforward method to place $m = k^2$ relay nodes is to position them in a regular $k \times k$ grid inside the square; see Fig. 7(a).

For a more sophisticated approach, one notes that in order to find an optimal placement for a set of relay nodes within a given sensor network, the locations of all the relay nodes should be considered at the same time. We, however, try to approximate the optimal solution by placing relay nodes into the network one at a time.

The algorithm performs a multidimensional search [2] in the following manner. Given a starting point y, a suitable direction d is first determined, and then the flow problem is optimized in this direction by performing a line search. Thereafter, a new direction d' is chosen and, again, the flow problem is optimized starting from the previous optimum in the direction d'. The process is repeated until a good enough solution is found, or the algorithm converges to a (possibly local) optimum.

In this case the optimizable quantity is the balanced data gathering measure F_λ, which can be calculated directly with the full LP model from Sect. 2, although requiring a large number of function evaluations. The optimal objective function value for the LP model with a given balancing factor λ and the considered relay node in location y is denoted $F_\lambda(y)$.

For our algorithm we have chosen as the starting point y^1 the center of mass of the differences of the offered data rates s_i and achieved data rates r_i:

$$y^1 = \left(\frac{\sum x^i{}_1(s_i - r_i)}{\sum (s_i - r_i)}, \frac{\sum x^i{}_2(s_i - r_i)}{\sum (s_i - r_i)} \right),$$

where $x^i = (x^i{}_1, x^i{}_2)$ are the coordinates of the sensor nodes in S.

Initialization Step Choose the number of iterations M and let $x^1 = (x^1{}_1, x^1{}_2)$ be
    the location of the sink node.
    Find the present optimal achieved data rates r_i by solving the corresponding
    linear program.
    Choose initial location for the new relay node as $y^1 = \left(\frac{\sum x^i{}_1 (s_i - r_i)}{\sum s_i - r_i}, \frac{\sum x^i{}_2 (s_i - r_i)}{\sum s_i - r_i} \right)$,
    let j=1, and go to the main step.
Main Step Repeat M times.
    1. Let $d^j = (y^j - x^1)$, let μ^j be a value that maximizes $F_\lambda(y^j + \mu d^j)$, and let
       $y^{j'} = y^j + \mu^j d^j$. Go to step 2.
    2. Let $d^{j'} \perp d^j$, let $\mu^{j'}$ be a value that maximizes $F_\lambda(y^{j'} + \mu^{j'} d_j')$, and let
       $y^{j+1} = y^{j'} + \mu^{j'} d_j'$.
    Increment j by one.

Fig. 6. The incremental relay node placement algorithm for placing one relay node.

The idea is to place a new relay node initially in a region of the network where the achieved data rates r_i are small compared to the offered rates s_i. It is reasonable to think that the ideal location of the node would be, at least with high probability, somewhere between this region and the sink. Therefore the first search line is chosen in direction of the sink node. This idea is extended for determining the remaining search directions as the algorithm proceeds. The search directions are chosen pairwise: in the direction of the sink node and orthogonal to it. Line searches can, in principle, be performed by almost any standard one-dimensional search method, the main limiting factors being the complexity and possible roughness of the objective function F_λ. Fig. 6 summarizes our algorithm for finding a good location for a relay node.

4.2   Experimental Results with Relay Nodes

The objective of the experiments was to analyze both the impact of relay nodes on the balanced data gathering measure and the performance of our incremental relay node placement algorithm. We used the obstacle-free sensor network given in Sect. 3 with balancing factor $\lambda = 0.5$. Relay nodes were assigned 100 times the energy of the sensor nodes (i.e., 2 kJ). The performance of our incremental algorithm was compared to the placement in a regular grid. The incremental algorithm was run using a uniform line search with 20 equidistant line points for each direction and with three different direction pairs ($M = 3$). An example of $m = 9$ relay nodes placed with the incremental algorithm is presented in Fig. 7(b). As can be seen from the figure, the relay nodes form a routing backbone and the sensor nodes exhibit a clustering behavior around the relay nodes.

The results for different numbers of relay nodes are shown in Fig. 7(c). A clear improvement in network performance can be seen with an increasing number of relay nodes, even with relatively simple relay node placement schemes. If the cost of a relay node is not considerably higher than the cost of a sensor node, augmenting sensor networks having tight energy constraints by relay nodes is worthwhile.

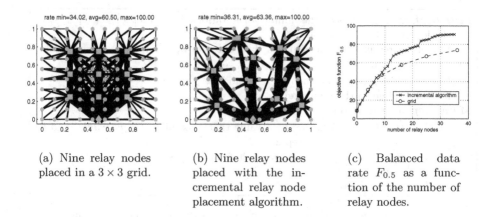

(a) Nine relay nodes placed in a 3 × 3 grid.

(b) Nine relay nodes placed with the incremental relay node placement algorithm.

(c) Balanced data rate $F_{0.5}$ as a function of the number of relay nodes.

Fig. 7. Experiments with 100 sensor nodes and relay nodes.

The incremental placement algorithm performs somewhat better than the grid placement algorithm, but is more demanding computationally. Unlike the incremental relay node placement algorithm, the straightforward grid placement of the relay nodes cannot be expected to perform as well for sensor networks where sensor nodes are placed arbitrarily, or where the area to be covered is irregular or partially obstructed.

5 Conclusions

We have considered the problem of energy-efficient data gathering in sensor networks, with special emphasis on the goal of balancing the average volume of data collected against sufficient coverage of the monitored area. We have formulated a linear programming model of the task of finding optimal routes for the data produced at the sensor nodes, given a balancing requirement in terms of a balancing factor $\lambda \in [0, 1]$.

Experiments with the model show that for reasonable values of the balancing factor, a significant increase in coverage is achieved, without any great decrease in the average amount of data gathered per node. In the examples considered in Sect. 3.1, for $\lambda = 0.5$, the minimum amount of data collected from any node was increased fourfold from the case $\lambda = 0$, with only about 10% decrease in the total volume of data gathered at the base station.

We have also considered the effects of augmenting a given network of sensors by a small number of auxiliary, freely positionable relay nodes with relatively high initial battery power levels. In the examples considered in Sect. 4.2, already an additional four relay nodes allocated in a network of 100 sensor nodes yielded a more than threefold increase in the value of the balanced data gathering objective function $F_{0.5}$; with nine relay nodes a fivefold increase was achieved. These improvements were obtained by a simple grid placement of the relay nodes; for larger numbers of relay nodes better results were achieved by an incremental relay placement heuristic, applying techniques of multidimensional line search.

In our experiments with obstacles, sensor networks were seen to be fairly robust against even a fairly high number of obstructions. This was achieved through the use of global optimization at a central location, where information about all link costs in the network was available. It remains to be studied how closely this global optimum can be approximated by distributed algorithms that have access to local information only. The effects of possible node faults during the operation of the network are also a topic for further research.

Acknowledgements. We would like to thank the anonymous referees for pointing out the recent publications [19,20] to us.

References

1. I. F. Akyildiz, W. Su, Y. Sankarasubramaniam, E. Cayirci, "Wireless sensor networks: a survey." *Computer Networks 38* (2002), 393–422.

2. M. S. Bazaraa, H. D. Sherali, C. M. Shetty, *Nonlinear Programming: Theory and Algorithms.* John Wiley & Sons, New York NY, 1993.

3. M. Bhardwaj, A. Chandrakasan, "Bounding the lifetime of sensor networks via optimal role assignment." *Proc. 21st Ann. Joint Conf. of the IEEE Computer and Communications Societies (InfoCom'02, New York NY, June 2002), Vol. 3*, 1587–1596. IEEE, New York NY, 2002.

4. M. Bhardwaj, T. Garnett, A. Chandrakasan, "Upper bounds on the lifetime of sensor networks." *Proc. IEEE Intl. Conf. on Communications (ICC'01, Helsinki, Finland, June 2001), Vol. 3*, 785–790. IEEE, New York NY, 2001.

5. A. Cerpa, J. Elson, D. Estrin, L. Girod, M. Hamilton, J. Zhao, "Habitat monitoring: application driver for wireless communication technology." *Proc. 2001 ACM SIGCOMM Workshop on Data Communications in Latin America and the Caribbean (San José, Costa Rica, April 2001)*, 20–41. ACM Press, New York NY, 2001.

6. K. Chakrabarty, S. S. Iyengar, H. Qi, E. Cho, "Grid coverage for surveillance and target location in distributed sensor networks." *IEEE Transactions on Computers 51* (2002), 1448–1453.

7. C.-Y. Chong, S. P. Kumar, "Sensor networks: evolution, opportunities, and challenges." In [22], pp. 1247–1256.

8. M. Chu, H. Haussecker, F. Zhao, "Scalable information-driven sensor querying and routing for ad hoc heterogeneous sensor networks." *Intl. J. on High Performance Computing Applications 16* (2002), 293–313.

9. *Communications of the ACM 43* (2000) 5 (May). Special section on Embedding the Internet.

10. *Communications of the ACM 45* (2002) 12 (December). Special section on Issues and Challenges in Ubiquitous Computing.

11. K. Dasgupta, M. Kukreja, K. Kalpakis, "Topology-aware placement and role assignment for energy-efficient information gathering in sensor networks." *Proc. 8th IEEE Symp. on Computers and Communication (ISCC'03, Kemer-Antalya, Turkey, July 2003)*, 341–348. IEEE, New York NY, 2003.

12. S. S. Dhillon, K. Chakrabarty, S. S. Iyengar, "Sensor placement for grid coverage under imprecise detections." *Proc. 5th ISIF Intl. Conf. on Information Fusion (FUSION'02, Annapolis MD, July 2002)*, 1581–1587. International Society for Information Fusion, Chatillon, France, 2002.

13. D. Estrin, R. Govindan, J. Heidemann, S. Kumar, "Next century challenges: scalable coordination in sensor networks." *ACM/IEEE Intl. Conf. on Mobile Computing and Networks (MobiCom'99, Seattle WA, August 1999)*, 263–270. ACM Press, New York NY, 1999.

14. D. Estrin et al., *Embedded, Everywhere: A Research Agenda for Networked Systems of Embedded Computers.* The National Academies Press, Washington DC, 2001.

15. W. R. Heinzelman, A. Chandrakasan, H. Balakrishnan, "Energy-efficient communication protocol for wireless microsensor networks." *Proc. 33rd Hawaii Intl. Conf. on System Sciences (HICSS'00, Maui HW, January 2000)*, 8020–8029. IEEE Computer Society, Los Alamitos CA, 2000.

16. A. Howard, M. J. Matarić, S. Sukhatme, "Mobile sensor network deployment using potential fields: a distributed, scalable solution to the area coverage problem." In H. Asama, T. Arai, T. Fukuda, T. Hasegawa (Eds.), *Distributed Autonomous Robotic Systems 5.* Springer-Verlag, Berlin, 2002.

17. K. Kalpakis, K. Dasgupta, P. Namjoshi, "Efficient algorithms for maximum lifetime data gathering and aggregation in wireless sensor networks." *Computer Networks* 42 (2003), 697–716.

18. B. Krishnamachari, D. Estrin, S. Wicker, "The impact of data aggregation in wireless sensor networks." *Intl. Workshop on Distributed Event-Based Systems (DEBS'02, Vienna, Austria, July 2002)*, 575–578. IEEE, New York NY, 2002.

19. B. Krishnamachari, F. Ordóñez, "Analysis of energy-efficient, fair routing in wireless sensor networks through non-linear optimization." *Proc. IEEE 58th Vehicular Technology Conference (VTC 2003-Fall, Orlando FL, October 2003)*, Vol. 5, 2844–2848. IEEE Computer Society, Los Alamitos CA, 2003.

20. F. Ordóñez and B. Krishnamachari, "Optimal information extraction in energy-limited wireless sensor networks." To appear in *IEEE JSAC, special issue on Fundamental Performance Limits of Wireless Sensor Networks*, 2004.

21. J. Pan, Y. T. Hou, L. Cai, Y. Shi, S. X. Shen, "Topology control for wireless sensor networks." *Proc. 9th Ann. Intl. Conf. on Mobile Computing and Networking (MobiCom'03)*, 286–299. ACM Press, New York NY, 2003.

22. *Proceedings of the IEEE 91* (2003) 8 (August). Special Issue on Sensor Networks and Applications.

23. T. S. Rappaport, *Wireless Communications: Principles & Practice.* Prentice Hall, Upper Saddle River NJ, 1996.

24. N. Sadagopan, B. Krishnamachari, "Maximizing data extraction in energy-limited sensor networks." *Proc. 23rd Conf. of the IEEE Computer and Communications Society (InfoCom'04, Hong Kong, March 2004)*, to appear.

Scale Free Aggregation in Sensor Networks*

Mihaela Enachescu[1], Ashish Goel[1], Ramesh Govindan[2], and Rajeev Motwani[1]

[1] Stanford University, Stanford, CA 94305
{mihaela,agoel,rajeev} @cs.stanford.edu
[2] University of Southern California, Los Angeles, CA 90089
ramesh@usc.edu

Abstract. Sensor networks are distributed data collection systems, frequently used for monitoring environments in which "nearby" data has a high degree of correlation. This induces opportunities for data aggregation, that are crucial given the severe energy constraints of the sensors. Thus it is very desirable to take advantage of data correlations in order to avoid transmitting redundancy. In our model, we formalize a notion of correlation, that can vary according to a parameter k. We propose a very simple randomized algorithm for routing information on a grid of sensors in a way which promotes data aggregation. We prove that this simple scheme is a constant factor approximation (in expectation) to the optimum aggregation tree *simultaneously* for all correlation parameters k. The key idea of our randomized analysis is to relate the expected collision time of random walks on the grid to scale free aggregation.

1 Introduction

Consider a network where each node gathers information from its vicinity and sends this information to a centralized processing agent. If the information is geographically correlated, then a large saving in data transmission costs may be obtained by aggregating information from nearby nodes before sending it to the central agent. This is particularly relevant to sensor networks where battery limitations dictate that data transmission be kept to a minimum, and where sensed data is often geographically correlated. In-network aggregation for sensor networks has been extensively studied over the last few years [9,7,14]. In this paper we show that a very simple opportunistic aggregation scheme can result in near-optimum performance under widely varying (and unknown) scales of correlation.

More formally, we consider the idealized setting where sensors are arranged on an $N \times N$ grid, and the centralized processing agent is located at position $(0, 0)$ on the grid. We assume that each sensor can communicate only to its four neighbors on the grid. This idealized setting has been widely used to study broad information processing issues in sensor networks (see [12], for example). We call

---

* Research supported in part by NSF grants CCR-0126347, IIS-0118173, EIA-0137761, and ITR-0331640, NSF Career grants No. 0339262 and No. 0339262, and a SNRC grant.

an aggregation scheme *opportunistic* if data from a sensor to the central agent is always sent over a shortest path, i.e., no extra routing penalty is incurred to achieve aggregation.

To model geographic correlations, we assume that each sensor can gather information in a $\frac{k}{2} \times \frac{k}{2}$ square (or, a circle of radius $k/2$) centered at the sensor. We will refer to k as the correlation parameter. Let the set $A(x)$ denote the area sensed by sensor i. If we aggregate information from a set of sensors S then the size of the resulting compressed information is $I(S) = \left|\bigcup_{x \in S} A(x)\right|$, i.e., the size of the total area covered by the sensors in S. Often, the parameter k can depend on the intensity of the information being sensed. For example, a volcanic eruption might be recorded by many more sensors and would correspond to a much higher k than a campfire. Accordingly, we will assume that the parameter k is not known in advance. In fact, we would like our opportunistic aggregation algorithms to work well simultaneously for all k.

There are scenarios where the above model applies directly. For example, the sensors could be cameras which take pictures within a certain radius, or they could be sensing RFID tags on retail items (or on birds which have been tagged for environmental monitoring) within a certain radius. Also, since we want algorithms that work well without any knowledge of k, our model applies to scenarios where the likelihood of sensing decreases with distance. For example, consider the case where a sensor can sense an event at distance r only if it has "intensity" $f(r)$ or larger, where f is a non-decreasing function. Then, events of intensity y correspond to information with correlation parameter roughly $f^{-1}(y)$; if these events are spread uniformly across the sensor field then an algorithm which works well for all k will also work well for this case.

Thus, we believe that our model (optimizing simultaneously for all k) captures the joint entropy of correlated sets of sensors in a natural way for a large variety of applications, a problem raised by Pattem *et al.* [12].

For node (i, j), we will refer to nodes $(i-1, j)$ and $(i, j-1)$ as its downstream neighbors, and nodes $(i+1, j)$ and $(i, j+1)$ as its upstream neighbors. We would like to construct a tree over which information flows to the central agent, and gets aggregated along the way. Since we are restricted to routing over shortest paths, each node has just one choice: which downstream node to choose as its parent in the tree. In our algorithm, a node (i, j) waits till both its upstream neighbors have sent their information out[1]. Then it aggregates the information it sensed locally with any information is received from its upstream neighbors and sends it on to one of its downstream neighbors. The cost of the tree is the total amount of (compressed) information sent out over links in the tree.

Note that we do not need all sensors at a certain distance to transmit synchronously; we just need to make sure that a node sends its information only after both its upstream nodes have transmitted theirs. This can be enforced asynchronously by each sensor. In any case, Madden et al [10] have developed

---

[1] Of course maybe one, or both, of the upstream nodes may decide not to chose (i, j) as the parent node; however we assume that node (i, j) gets notified anyway when its upstream nodes send information out.

protocols to facilitate synchronous sending of information by sensors (depending on the distance from the sink) which we can leverage if needed.

Our Results: We propose a very simple randomized algorithm for choosing the next neighbor – node (i, j) chooses its left neighbor with probability $i/(i+j)$ and its bottom neighbor with probability $j/(i + j)$. Observe that this scheme results in all shortest paths between (i, j) and $(0, 0)$ being chosen with equal probability. We prove that this simple scheme is a constant factor approximation (in expectation) to the optimum aggregation tree *simultaneously* for all correlation parameters k. While we construct a single tree, the optimum trees for different correlation parameters may be different.

The key idea is to relate the expected collision time of random walks on the grid to scale free aggregation. Consider two neighboring nodes X and Y, and randomly choose a shortest path from each of the nodes to the sink. Define the collision time to be the number of hops (starting at X or Y) before the paths first meet. We first show (Sect. 3) that if the average expected collision time is $O\left(\sqrt{N}\right)$, then we have a constant factor approximation algorithm to the optimal aggregation for *all* correlation parameters k. We then show that the average expected collision time for our randomized algorithm is indeed $O\left(\sqrt{N}\right)$ (Sect. 4). This analysis of the average expected collision time is our main technical theorem and may be of independent interest. To achieve this result, we first analyze the expected number of differing steps (where the two paths move in different directions) and then prove that the probability of a step being a differing step is a super-martingale.

We also present (Sect. 5) a slightly more involved hierarchical routing algorithm that is deterministic, and has an average collision time of only $O(\log N)$; hence the deterministic algorithm is also a constant factor approximation for all correlation parameters k. While this scheme has a slightly better performance, we believe that the simplicity of the randomized algorithm makes it more useful from a practical point of view.

Our results hold only for the total cost, and critically rely on the fact that information is distributed evenly through the sensor field. It is easy to construct pathological cases where our algorithm will not result in good aggregation if information is selectively placed in adversarially chosen locations.

This result shows that, at least for the class of aggregation functions and the grid topology considered in this paper, schemes that attempt to construct specialized routing structures in order to improve the likelihood of data aggregation [6] are unnecessary. This is convenient, since such specialized routing structures are hard to build without some a priori knowledge about correlations in the data. With this result, simple geographic routing schemes like GPSR [8], or tree-based data gathering protocols are sufficient [7,10].

Related Work: Given the severe energy constraints and high transmission cost in the sensor network setting, data aggregation has been recognized as a crucial operation, which optimizes performance and longevity [4]. In the sensor network literature, aggregation can refer to either a database aggregate operator (min,max,sum etc.) as in [1,10,11], or to general aggregation functions such as the one we consider in this paper.

Goel and Estrin [3] studied routing that leads to a simultaneously good solution (a $\log n$ approximation) for a large class of aggregation functions. In their model, the amount of aggregation only depends on the number of nodes involved, independent of location. In our problem, the amount of aggregation depends on the location of the sensors being aggregated: the closer two sensors are, the more correlated their data is. But it is also easier to aggregate data from nearby nodes. Hence, it seems intuitive that better simultaneous optimization may be possible for our case, an intuition that we have verified in this paper.

We build on the work of Pattem et al. [12] who study a closely related question, comparing three different classes of compression schemes for sensor networks: routing-driven compression, in which the nodes just follow a shortest path and compression is done opportunisticaly whenever possible, compression-driven routing which builds up a specialized routing structure, and distributed source coding which leverages a priori information about correlations. After a theoretical analysis of these schemes, they introduce a generalized cluster-based compression scheme in which correlated readings are aggregated at a cluster head, which is studied via simulations. They find that across a wide variety of correlations (roughly parameterized by the joint entropy of two sensors spaced d apart), the cluster-based compression scheme works reasonably well with a relatively fixed cluster size. Our model captures a wider range of joint entropy functions (since we also approximate any linear composition of k-correlated information for different values of k), one of the open problems they pose. Also, we present a formal proof of simultaneous optimization. It is easy to see that their cluster-based compression scheme does not perform well in our model, in that no single cluster size can be within constant factor of the optimal aggregation tree for all k.

A study of coding schemes for correlated sensors has been performed by [2].

Before presenting our algorithms and analysis, we define the problem more formally.

2   Problem Definition

We consider the idealized setting where sensors are arranged in a $N \times N$ grid. There is a centralized processing agent at $(0,0)$. Each sensor can only communicate with its four nearest neighbors. The sensor network can sense multiple kinds of data.

For a specific type of data, we will refer to the information contained in a 1×1 grid square as a *value*. We then define this type of data k-correlated data if the following holds:

(i) Each value is sensed by all the sensors in a $k \times k$ grid centered at that location, as in Fig. 1. We will assume for simplicity that k is even, so that the notion of centering is well defined.

(ii) Let $A_k(x)$ denote the set of grid squares sensed by sensor x. If information from a set S of sensors is aggregated, the resulting information is of size $\left| \bigcup_{x \in S} A_k(x) \right|$

Fig. 1. The $k \times k$ sensors that detect a given value (the black square) for $k = 4$

We will look at k-correlated data for which $k < N/2$, since otherwise we obtain a pathological case in which all information can be captured by a single sensor in the network.

We want to find a tree on which to send information from all sensors to the center so as to approximately minimize the cost, simultaneously, for all values of k. In our cost model we focus on the transmission cost: each time a value is transmitted from a node to a neighbor the total cost increases by 1.

We can assume that each sensor knows it's (x,y)-coordinates. This can be done via the fine-grain localization method described by Savvides et. al in [13].

We also assume that a sensor cannot withhold information, and needs to send all information it can sense. The transmission is synchronized so that a node sends information only after receiving all data from its upstream neighbors, and finishes aggregating that information with its own data.

Theorem 1. *Lower bound on optimum cost (OPT) is $\theta\left(N^3 + (Nk)^2\right)$.*

Proof. Look at one individual value, at point (x, y) with $x, y \geq 0$ and construct the minimum cost routing for it.

The closest node to the origin that senses this value, at coordinates $(x - k/2, y - k/2)$, has to send the value all the way to the origin, so a cost of $D = x + y - k$ (distance from the point to the origin must be paid). All values incur this cost.

The value must be transmitted by all the nodes that can sense it, each node thus introducing a cost of 1. Thus, for all values for which the sensing $k \times k$ square is included in the $N \times N$ grid (it is easy to observe that there are $(N - k)^2$ such values), there is a cost of at least k^2 before the distinct values can be aggregated at a single node. We ignore this contribution to the cost for the other values.

Since we assume different values cannot be aggregated between them, we get a lower bound for the overall cost of at least:

$$\sum_{\text{values}} D + (N-k)^2 k^2 = 4 \left(\sum_{0<x<N, 0<y<N} (x+y-k) \right) + (k(N-k))^2 =$$

$$= N^2(N-1-k) + (k(N-k))^2$$

If we only consider parameters $k < N/2$ then $N-1-k \leq N/2$ and the above becomes $\theta\left(N^3 + (Nk)^2\right)$ as desired.

Note that this solution may not be feasible, because, a given sensor, from the point of view of one value should maybe communicate with one downstream neighbor for optimal aggregation, while from the point of view of another value should communicate with the other downstream neighbor, leading to an impossible solution for the "aggregation tree". However this analysis definitely gives a lower bound on OPT.

3   Relating Opportunistic Aggregation to Collision Time

Consider any algorithm which sends the information from node X to the center on a shortest path, and call such an algorithm opportunistic. The routing tree induces a path from any node to the center. The paths from two neighbors X and Y will eventually meet at some point Z. We call the distance from X to Z the collision time of X and Y. We will show the following theorem:

Theorem 2. *An opportunistic algorithm with average expected collision time* $O\left(\sqrt{N}\right)$ *gives a constant factor approximation to the optimum aggregation algorithm for all k.*

Proof. In a similar fashion as in our proof for the lower bound for the OPT, look from the point of view of a data value which is shared by $k \times k$ sensors. Consider the left and lower sides of the $k \times k$ sensing square. The paths from all sensors inside the square will go through one of the points on the sides.

Inside the square we pay the same cost as in our lower bound for the OPT (i.e. at most k^2). We now need to consider only the paths from the sides of the square to $(0,0)$. The distance before two adjacent paths meet introduces extra cost (two instances of the same info are transmitted, as opposed to only one instance, as would happen in the optimal case). It is easy to see that for each pair of adjacent nodes, there are k values that incur the extra cost due to the collision time.

Thus, the cost of our algorithm is given by:

$$\sum_{\text{values}} D + k^2 N^2 + k \sum_{\text{sensors}} (\text{collision time of the paths from two adjacent sensors}) =$$

$$= \theta\left(N^3 + (kN)^2\right) + O\left(kN^{2.5}\right)$$

The first two terms are the same as in the lower bound for OPT.

If $k < \sqrt{N}$ then the N^3 term dominates the $(Nk)^2$ term, as well as the $k \times N^{2.5}$ term, and we get an $O(1)$-approximation.

If $k > \sqrt{N}$ then the $(Nk)^2$ term dominates the other two terms, and we get again an $O(1)$-approximation.

Note that we compare to a lower bound for OPT, not OPT itself, which may be hard to compute, so the constant factor would be even less than what we can compute here.

4   The Probabilistic Distribution Shortest Path Algorithm

We will present a simple randomized opportunistic algorithm for constructing a tree. The path from each node will be a random walk towards the center, but the walks are not independent. The main result is to prove that the average expected collision time of two adjacent paths is $O\left(\sqrt{N}\right)$. The analysis of our random process may well be of independent interest.

Then, applying theorem 2 we can conclude this algorithm produces a constant factor approximation of the optimal aggregation trees for any value of k.

The Probabilistic Distribution Algorithm: For every node, if the node is located at position (x, y) choose to include in the MST the left-edge with probability $\frac{x}{x+y}$ and the down-edge with probability $\frac{y}{x+y}$.

The Random Walk view: We can view the above process as a tree constructed from random walks originating from each sensor. At each time step the current node chooses one of the (at most) two downstream nodes as its parent. Because a node waits for its upstream nodes to transmit we can view the process as a flow in which the data gets closer by one to the origin at each time step. In our model, when two walks meet (passing some step through the same node) they "collapse" into a single walk and lose their independence. The analysis of the expected collision time for this random process is presented below.

4.1   Proving the Average Collision Time of the Random Walks

Theorem 3 (Random Walk Theorem). *The average expected collision time of two adjacent walks as generated by the randomized probabilistic distribution algorithm is $O\left(\sqrt{N}\right)$.*

Let us first introduce some notation, definitions, and lemmas which together will imply the above result.

Two neighboring nodes can be either horizontal or vertical neighbors, and one, say the second, is the upstream neighbor of the other. Thus there is a $\frac{x}{x+y}$ probability to meet initially. If they do not meet initially, then the upstream node chooses as its parent another node, which on the grid is at distance 2 from the first node, and at the same distance from the origin. Let's assume that the two walks do not meet initially.

We will analyze the collision time of the random walks originating at $(x-1, y)$ and $(x, y-1)$. Because the nodes are at the same distance from the origin they move towards the central point in sink. Look at the horizontal difference between the two paths, as a function of time, and let's denote this by $\Delta_t(x, y)$. Initially, $\Delta_0(x, y) = 1$. Because in general we focus our attention to a specific (x, y) we will drop these parameters from the notation. We want to analyze $E[t_c]$ where t_c is such that $\Delta_t = 0$ for the first time. Observe that t_c is precisely the collision time as defined earlier, since the two walks start from the same distance from the origin, and at every time step we assume the walks move one unit closer, so there is a one-to-one correspondence between time and distance from the initial point to the collision point. Once the horizontal distances become equal, the vertical distances must also be equal and the two paths would meet.

Let $M = x + y - 1$, the initial distance from the origin.

At each time step, Δ_t can stay the same or become different (increase or decrease by 1). We call a step at which Δ_t differs from Δ_{t-1} a *differing* step.

We will first analyze the number of differing steps before collision, and then bound the probability that a step is a differing step.

Definition 1. *Let's denote by $D(x, y)$ the number of differing steps before Δ_t becomes 0 for the first time.*

Lemma 1. $E[D(x, y)]$ *is* $O\left(\sqrt{min(x, y)}\right)$.

Proof. At time t, when the first path is at, say, point (x_1, y_1) and the second path is at point (x_2, y_2) we know that $x_1 + y_1 = x_2 + y_2 = M - t$. Initially $x_1 < x_2$, so this will continue to hold until $\Delta = x_2 - x_1$ first becomes 0. Also, initially, $x_2 = x$ and $y_1 = y$.

Based on our probabilistic model, for the next time step we have:

$$\Pr(\Delta_{t+1} - \Delta_t = 1) = \frac{x_1 y_2}{(x_1 + y_1)^2} \text{ and } \Pr(\Delta_{t+1} - \Delta_t = -1) = \frac{x_2 y_1}{(x_1 + y_1)^2}$$

Using $y_1 = M - t - x_1$ and $y_2 = M - t - x_2$ we obtain the following:

$$\Pr(\Delta_{t+1} - \Delta_t = 1) + \Pr(\Delta_{t+1} - \Delta_t = -1) = \frac{(M - t)(x_1 + x_2) - 2x_1 x_2}{(M - t)^2}$$

and $\Pr(\Delta_{t+1} - \Delta_t = 1) - P(\Delta_{t+1} - \Delta_t = -1) = -\frac{\Delta_t}{M - t}$

Now define $p_f(t) = P(\Delta_{t+1} - \Delta_t = 1 | \Delta_{t+1} - \Delta_t \neq 0)$ and $p_r(t) = P(\Delta_{t+1} - \Delta_t = -1 | \Delta_{t+1} - \Delta_t \neq 0)$ to be the conditional (normalized) probabilities of a forward (positive) change in Δ, and of a reverse (negative) change in Δ, respectively. Also define λ as below:

$$\lambda = p_f(t) - p_r(t) = \frac{P(\Delta_{t+1} - \Delta_t = 1) - P(\Delta_{t+1} - \Delta_t = -1)}{P(\Delta_{t+1} - \Delta_t = 1) + P(\Delta_{t+1} - \Delta_t = 1)} = \frac{\Delta_t(M - t)}{(M - t)(x_1 + x_2) - 2x_1 x_2}$$

Since $p_f(t) + p_r(t) = 1$, we can rewrite $p_f(t)$ and $p_r(t)$ as:

$$p_f(t) = \frac{1}{2} - \frac{\lambda}{2} \text{ and } p_r(t) = \frac{1}{2} + \frac{\lambda}{2}$$

where λ still contains a dependence on t. The convergence to $\Delta = 0$ can only be slower if λ is smaller. Note that by removing the $2x_1 x_2$ term from the denominator of λ we can only decrease the overall value of λ. Also, we get the same effect if we replace $x_1 + x_2$ by $2\max(x_1, x_2) = 2x_2$

Also, the $M - t$ factor will get simplified so we can replace λ by $\frac{\Delta}{2x_2}$ to obtain new forward and reverse probabilities, independent of t and only dependent on Δ:

$$n_f(\Delta) = \frac{1}{2} - \frac{\Delta}{4x_2} \text{ and } n_r(\Delta) = \frac{1}{2} - \frac{\Delta}{4x_2}$$

Now consider an integer random walk in $[0, \max(x_1, x_2) = x_2]$, with an absorbing barrier at 0, and a reflecting one at $\max(x_1, x_2) = x_2$.

We analyze the behavior of this random walk in lemma 2. By construction, the expected time for this new random walk to reach 0 starting from 1 is an upper bound to the expected time for Δ to reach 0 starting from 1.

We can then conclude that Δ reaches 0 in $O\left(\sqrt{x_2}\right)$ by directly applying the result in lemma 2. By symmetry we can also obtain time $O\left(\sqrt{y_1}\right)$. Since $x_2 = x$ and $y_1 = y$ initially, the theorem is proven.

Lemma 2. *Consider an integer random walk starting at point 1 on the interval $[0, x]$. Assume that, if we are at position j the random walk moves right with probability $n_f(j)$, and left with probability $n_r(j)$ in the interval $[1, x - 1]$ where $n_f(j)$ and $n_r(j)$ are as defined in lemma 1. Assume that the point 0 is absorbing, and that the point x is reflecting (i.e. the walk moves to $x - 1$ with probability 1 from x). If the walk starts at point 1, then the expected number of time steps necessary for this walk to first reach 0 is $O\left(\sqrt{x}\right)$.*

Proof. Note that at each step we move either in one direction or the other, since, by definition, $n_r + n_f = 1$

Define $B(j)$ to be the expected number of steps before the point $j - 1$ is first visited, assuming that the random walk starts at point i. We are then looking for the value of $B(1)$. We will use the properties of the walk, in particular the values of $n_f(j)$ and $n_r(j)$ to derive a recursive formula for $B(j)$ and then get a bound for $B(1)$.

If we pass exactly $i + 1$ times through point j before reaching point $j - 1$, the expected number of steps is $iB(j + 1) + 1$. The probability of this event is $n_r(j)n_f(j)^i$. Since i can range from 0 to ∞ we get the following relation for $B(j)$, where $j \in [1, x - 1]$:

$$B(j) = \sum_{i=0}^{\infty} n_r(j)n_f(j)^i(iB(j+1)+1) = n_r(j)\sum_{i=0}^{\infty} n_f(j)^i + n_r(j)B(j+1)\sum_{i=0}^{\infty} n_f(j)^i i$$

$$= \frac{n_r(j)}{1 - n_f(j)} + \frac{n_r(j)n_f(j)}{(1 - n_f(j))^2} = 1 + \frac{n_f(j)}{n_r(j)}B(j+1) = \frac{2x-j}{2x+j}B(j+1) + 1$$

Further note that $B(x) = 1$ because x is a reflecting barrier, so in the next step we move back with probability 1.

We want to solve for $B(1)$, the value of interest.

If we expand $B(1)$ in terms of $B(x)$ we obtain:

$$B(1) = \sum_{i=1}^{2d} \frac{2x-1}{2x+1} \times \ldots \times \frac{2x-i}{2x+i}$$

To simplify notation, denote $2x$ by X and $\frac{2x-1}{2x+1} \times \ldots \times \frac{2x-i}{2x+i}$ by T_i.

Now, note that for $i \in \{1, \ldots, 2\sqrt{D}\}$ we have $T_i \leq 1$.

For $i \in \{2\sqrt{X} + 1, \ldots, 3\sqrt{X}\}$ we have $T_i \leq \left(\frac{X-\sqrt{X}}{X+\sqrt{X}}\right)^{\sqrt{X}}$.

In general, for any m, if $i \in \{m\sqrt{X} + 1, \ldots, (m+1)\sqrt{X}\}$ we have $T_i \leq \left(\frac{X-\sqrt{X}}{X+\sqrt{X}}\right)^{\sqrt{mX}}$.

Thus $B(1)$ can be upper bounded by a geometric series with sum $\frac{X+\sqrt{X}}{2\sqrt{X}}$.

Note that the $\left(\frac{X-\sqrt{X}}{X+\sqrt{X}}\right)^{\sqrt{X}}$ is approximately e^2, and thus constant, for large enough X, where $X = 2x$. Thus, the first term (the fraction) of this bound is a constant, and we conclude that $B(1)$ is $O(\sqrt{x})$.

Definition 2. *Define $p_t(x, y) = \Pr[\Delta_t$ is differing $\mid$ two walks have not collided yet$]$.*

As before, we will omit the arguments x, y since they are fixed.

Lemma 3. *For all t, $p_{t+1} \geq p_t$*

Proof. Suppose the first walk is at coordinates (i, j) and the second one at co-ordinates $(i + \Delta_t, j - \Delta_t)$.

Case 1 ($\Delta_t \geq 2$): Then $\Delta_{t+1} \geq 1$, since the difference between Δ_t and Δ_{t+1} can be at most 1. Thus the random walks would not meet at time $t+1$, so we eliminate the conditioning for p_{t+1}, and we have the following:

$$\Pr[\Delta_t \text{ is differing}] = f(i, j, \Delta_t) = \frac{i(j - \Delta_t) + j(i + \Delta_t)}{(i + j)^2}$$

$$\Pr[\Delta_{t+1} \text{ is differing}] = g(i, j, \Delta_t) = \frac{i(j - \Delta_t)f(i - 1, j, \Delta_t + 1)}{(i + j)^2} +$$

$$+ \frac{i(i + \Delta_t)f(i - 1, j, \Delta_t)}{(i + j)^2} + \frac{j(i + \Delta_t)f(i, j - 1, \Delta_t - 1)}{(i + j)^2} + \frac{j(j - \Delta_t)f(i, j - 1, \Delta_t)}{(i + j)^2}$$

It is easy to verify, using Mathematica for example, that $f(i, j, \Delta_t) - g(i, j, \Delta_t) = 0$, and hence, $p_t = p_{t+1}$.

Case 2 ($\Delta_t = 1$): In this case the conditioning in the definition of p_{t+1} implies that one of the cases in the above formula cannot take place. We still obtain that $p_{t+1} \geq p_t$, but the details are technical and are deferred to the appendix.

Proof (Random Walk Theorem).

From Lemma 1 we extract the number of differing steps before two neighboring walks meet.

Consider two walks at (x, y) and $(x + 1, y - 1)$.

From lemma 3 we see that at each time step the probability of a differing step is bounded below by p_0, the initial probability of having a differing step, which is given by

$$\frac{x(y - 1) + (x + 1)y}{(x + y)^2} = \theta\left(\min(x, y)\frac{2\max(x, y)}{(x + y)^2}\right) = \theta\left(\frac{\min(x, y)}{x + y}\right)$$

Thus the expected time before a differing step is $\theta\left(\frac{x+y}{\min(x,y)}\right)$.

Combining this with result from Lemma 1 which bounds the expected number of differing steps before the two walks meet we obtain:

$$E[\text{collision time for neighboring walks at } (x, y)] = \theta\left(\sqrt{\min(x, y)}\frac{x+y}{\min(x,y)}\right)$$

$$= \theta\left(\frac{x+y}{\sqrt{\min(x,y)}}\right)$$

Taking the sum over all x, y pairs we obtain: $\sum_{x,y}\left(\frac{x+y}{\sqrt{\min(x,y)}}\right) = \theta(N^{2.5})$

which concludes the proof of the expected average collision time theorem.

5   The Hierarchical Decomposition Approach

We now present a deterministic algorithm for constructing a tree that produces a constant factor approximation for any value of k. This algorithm has better properties (its average collision time is $O(\log N)$ instead of $O\left(\sqrt{N}\right)$ for example), but it is more involved. Also the approximation provided is still $O(1)$.

The solution is based on the idea of dividing the grid into sub-grids, and collecting all the values in a given sub-grid at the sensor closest to the origin before forwarding it onto the next sub-grid.

5.1   The Hierarchical Decomposition Algorithm

We present the construction and the proof of correctness in parallel. We need two stages: a top-down stage in which we establish the sub-grids recursively, and a bottom-up stage in which we put the sub-grids together. We will assume for simplicity that N is a power of 2.

The Top-Down Stage. Divide the first quadrant in four sub-grids of size $N/2 \times N/2$, each of which is further divided in four size-$N/4 \times N/4$ sub-grids, and so on. For each sub-grid we will make sure that the MST converges to the sensor closest to the origin, i.e. if there is choice in what direction to move towards the origin, choose the choice that would not leave the sub-grid. If there is still choice choose arbitrarily.

The Bottom-Up Stage. We will prove by construction the following lemma.

Lemma 4. *If a $2^k \times 2^k$ sub-grid has the property that its average collision time is less than ck for all adjacent node pairs in the sub-grid, then we can construct a $2^{k+1} \times 2^{k+1}$ sub-grid with average collision time of $c(k+1)$ for all adjacent node pairs, where c is some constant greater than 2.*

Proof. We assume the parent node is determined for all nodes inside the $2^k \times 2^k$ sub-grid, and thus we have constructed an MST, rooted at the sensor node closest to the origin, such that the property is true. If we combine four copies of this construction, as in Fig. 2 we need to establish the parent node of the three root sensors B, D, and C representing the upper-left, lower-right, and upper right sub-grids respectively. For the first two the choice is forced (the sensor at B needs to go left, and the one at D needs to go down). For the third (the sensor at C) let us route to the left.

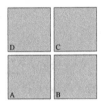

Fig. 2. Combining 4 smaller sub-grids to create the sub-grid at the next level

Now calculate the new average for the $2^{k+1} \times 2^{k+1}$ sub-grid, assuming the hypothesis holds for the $2^k \times 2^k$ ones.

We have $2(2^k)^2$ pairs included in each of the 4 smaller sub-grids, and thus have average less then ck, from the hypothesis. We also have 2^{k+2} new pairs (the ones spanning the white lines) that have collision time bounded by 2^{k+2}. Thus we obtain a new average collision time of: $\frac{8ck(2^k)^2+(2^{k+2})^2}{2(2^{k+1})^2} \leq ck+2 \leq c(k+1)$
as long as $c > 2$.

The base case is trivial.

6 Conclusions and Future Work

In this paper, we have argued that there exists a routing tree which is a constant factor approximation (in expectation) to the optimum aggregation tree *simultaneously* for all correlation parameters k. We present two constructions and prove that they obtain a constant approximation factor. Our result has important consequences – it obviates the need for specialized routing structures at least for the class of aggregation functions considered in this paper. This is convenient, since such specialized routing structures are hard to build without some a priori knowledge about correlations in the data.

There are several possible future research directions that this work leads to. It would be interesting to study the behavior of our randomized algorithm for non-grid topologies (for example on a random graph), or for the grid-topology model with generalized connectivity assumption, in which nodes have a larger number of neighbors. Another research direction would be to extend the aggregation model, either by defining a more general framework, or by analyzing the range of aggregation functions that can be obtained by combining the already defined functions.

References

[1] P. Bonnet, J. Gehrke, and P. Seshadri, Querying the Physical World, *IEEE Personal Communications Special Issue on Networking the Physical World*, October 2000.

[2] R. Cristescu, B. Beferull-Lozano, and M. Vetterli, On Network Correlated Data Gathering, *IEEE Proceedings of INFOCOM*, 2004, Hong Kong.

[3] A. Goel and D. Estrin, Simultaneous optimization for concave costs: single sink aggregation or single source buy-at-bulk, *Fourteenth Annual ACM-SIAM Symposium on Discrete Algorithms (SODA)*, 2003, p. 499-505.

[4] J. Heidemann, F. Silva, C. Intanagonwiwat, R. Govindan, D. Estrin, and D. Ganesan, Building Efficient Wireless Sensor Networks with Low-Level Naming, *Symposium on Operating Systems Principles*, 2001.

[5] W. Heinzelman, A. Chandrakasan, and H. Balakrishnan, Energy-Ecient Communication Protocol for Wireless Microsensor Networks, *33rd Hawaii International Conference on System Sciences (HICSS '00)*, 2000.

[6] C. Intanagonwiwat, D. Estrin, R. Govindan, and J. Heidemann, The Impact of Network Density on Data Aggregation in Wireless Sensor Networks, *ICDCS*, 2002.

[7] C. Intanagonwiwat, R. Govindan, D. Estrin, J. S. Heidemann, and F. Silva, Directed diffusion for wireless sensor networking, *IEEE/ACM Transactions on Networking*, vol. 11, no. 1, p. 2-16, 2003.

[8] B. Karp and H. T. Kung, GPSR: Greedy Perimeter Stateless Routing for Wireless Networks, *Mobile Computing and Networking (MobiCom)*, 2000.

[9] B. Krishnamachari, D. Estrin, and S. B. Wicker, The Impact of Data Aggregation in Wireless Sensor Networks, *ICDCS Workshop on Distributed Event-based Systems (DEBS)*, 2002.

[10] S. R. Madden, M. J. Franklin, J. M. Hellerstein, and W. Hong, TAG: a Tiny AGgregation Service for Ad-Hoc Sensor Networks, *Fifth Annual Symposium on Operating Systems Design and Implementation (OSDI)*, 2002.

[11] S. R. Madden, R. Szewczyk, M. J. Franklin, and D. Culler, Supporting Aggregate Queries Over Ad-Hoc Wireless Sensor Networks, *Fourth IEEE Workshop on Mobile Computing and Systems Applications*, 2002.

[12] S. Pattem, B. Krishnmachari, and R. Govindan, The Impact of Spatial Correlation on Routing with Compression in Wireless Sensor Networks, *Symposium on Information Processing in Sensor Networks (IPSN)*, 2004.

[13] A. Savvides, C.C. Han, and M. B. Strivastava, Dynamic Fine-Grained Localization in Ad-Hoc Networks of Sensors, *MobiCom*, 2001.

[14] A. Scaglione and S. D. Servetto, On the Interdependence of Routing and Data Compression in Multi-Hop Sensor Networks, *MobiCom*, 2002.

A   Technical Details for Case 2 of Lemma 3

Since we want to maintain $\Delta_{t+1} \geq 1$ (no collision at time $t + 1$), we eliminate the case in which the first walk moves from $(i, j - 1)$ to $(i - 1, j - 1)$ and the second walk moves from $i - 1, j$ to the same point as the first walk.

Thus our formula for p_{t+1} becomes:

$\Pr[\Delta_{t+1}$ is differing | the two walks do not collide $]$

$$= \frac{i(j - \Delta_t)f(i - 1, j, \Delta_t + 1)}{(i + j)^2} +$$

$$+ \frac{i(i + \Delta_t)f(i - 1, j, \Delta_t)}{(i + j)^2} \frac{j(j - \Delta_t)f(i, j - 1, \Delta_t)}{(i + j)^2}$$

$$= g(i, j, \Delta_t) - \frac{j(i + \Delta_t)f(i, j - 1, \Delta_t - 1)}{(i + j)^2}$$

while the formula for p_t remains

$$\Pr[\Delta_t \text{ is differing}] = f(i, j, \Delta_t) = \frac{i(j - \Delta_t) + j(i + \Delta_t)}{(i + j)^2}$$

Taking the difference between the two, and simplifying, using Mathematica for example, we obtain:

$$(p_{t+1} - p_t)(i + j)^2 = i^3 - i^2(j - 2) - i(j - 1)^2 + (j - 1)^2 j$$

If $j > i$ the right hand side reduces to $2i^2 + (j - i)[(j - 1)^2] - i^2$ which is positive; if $i > j$ the right hand side reduces to $[i^2 - j^2](i - j) + 2i$ which is again positive; if $i = j$ the right hand side is just $2i^2$, again positive.

Combining this with the fact that $(i + j)^2 \geq 0$ for all i, j we deduce that $p_{t+1} - p_t$ is always positive, which is exactly what we wanted to prove.

The Expected Uncertainty of Range Free Localization Protocols in Sensor Networks

Gideon Stupp and Moshe Sidi

EE Dept., Technion, Haifa 32000, Israel.
stupp@ee.technion.ac.il

Abstract. We consider three range-free localization protocols for sensor networks and analyze their accuracy in terms of the expected area of uncertainty of position per sensor. Assuming a small set of anchor nodes that know their position and broadcast it, we consider at first the simple Intersection protocol. In this protocol a sensor assumes its position is within the part of the plane that is covered by all the broadcasts it can receive. This protocol was also analyzed in [10], but our result contrast theirs. We then extend this protocol by assuming every sensor is preloaded with the entire arrangement of anchors before being deployed. We show that in this case the same expected uncertainty can be achieved with 1/2 the anchor nodes. Finally we propose an approximation for the arrangement based protocol which does not require any preliminary steps and prove that its accuracy converges to that of the arrangement protocol with high probability as the number of anchors increases.

1 Introduction

Sensor networks provide wireless connectivity for many stationary sensors that are usually randomly embedded in some physical domain. The sensors' purpose is to monitor and report locally sensed events so it is common to assume that a sensor can position itself within some global coordinate system. Since the standard GPS positioning is considered too costly for deployment within every sensor, other localization mechanisms specific for sensor networks are typically used. Such mechanisms can be divided into *range based* systems and *range free* systems.

Range based systems augment every sensor with hardware that is specifically necessary for localization. For example, augmented sensors may be able to deduce their distance from each other. This information is extremely useful and such systems can generally achieve fine grained localization. However, augmenting in this way the hardware of every sensor may be very costly. Range free systems, on the other hand, attempt to reduce the number of augmented sensors to a minimum. A small subset of the sensors is augmented, typically with full GPS positioning systems. These sensors (called *anchors*) broadcast their position and the other sensors use this information to localize themselves. The most important

S. Nikoletseas and J. Rolim (Eds.): ALGOSENSORS 2004, LNCS 3121, pp. 85–97, 2004.

success measure for such systems is the localization error given a small, fixed number of anchors.

Although many range free localization systems were suggested [7,6,2,9] in the past, not a lot has been done in terms of analytically evaluating their expected error. In this paper we consider three range free localization protocols starting with the basic intersection protocol [10], and analyze their expected accuracy in terms of the area of uncertainty of position per sensor. While making some simplifying assumptions for the analysis (we consider a discrete model where distances are in L_∞) the algorithms themselves are not dependent on the model. As we show, simple enhancements to the basic protocol produce significant reductions in the expected area of uncertainty.

Assume that N sensors are randomly placed in an $[n + 1] \times [n + 1]$ grid of congruent cells and that only a subset of size $K < N$ of these sensors (the *anchors*) know their position in terms of the coordinates of the cell they occupy. Have these anchors broadcast their position and let that broadcast be received only by sensors that are within the *communication range*, ρ, defined as the rectangle of size $[2\rho + 1] \times [2\rho + 1]$ centered at the broadcasting anchor; see Fig. 1(a). Denoting as *neighbors* sensors that can directly communicate with each other we describe three localization protocols. In the *Intersection* protocol, a sensor can position itself anywhere within the intersection of the communication ranges of its neighbor anchors; see Fig. 1(b). In the *Arrangement* protocol, a sensor can position itself at the intersection of its neighbor anchors that is not within range of any other anchor; see Fig. 1(c). Finally, in the *Approximate Arrangement* protocol, a sensor can position itself anywhere within range of its neighbor anchors that is out of range of its neighbor neighbors' anchors; see Fig. 1(d).

We define the *uncertainly* of position for a sensor to be the number of cells considered as possible locations for it and present analytic expressions for the expected uncertainty in the Intersection and Arrangement protocols. Furthermore, we show that the Arrangement protocol reaches the same expected uncertainty as the Intersection protocol with at most $1/2$ of the anchors. We then consider the approximation protocol, showing both analytically and by simulations that its expectation converges to that of the Arrangement protocol.

Related Work: Numerous range free localization mechanisms are described in the literature. The first solutions were typically central. For example, the connectivity matrix of the sensors is used in [3] to constraint the possible locations of the sensors in the network, achieving a position estimation by centrally solving a convex optimization problem. Many distributed solutions were also suggested. In the Centroid strategy, sensors locate themselves at the center of mass (the centroid) of the locations of anchors they can detect [2,1]. Other approaches [7, 6] assumed node to node communication can be used to flood the location of all the anchors to all the sensors. Using the hop-count as an estimate of the euclidian distance (by computing the average distance between sensors [7] or by analytically deriving it [6]), a sensor can estimate its position via triangulation.

The accuracy of all of these solutions, however, was evaluated experimentally, rather than analytically.

The research of range based systems has been more extensive, supplying bounds on the achievable positioning accuracy when assuming specific calibration data, noise patterns and location distribution. In particular, lower bounds for localization uncertainty using the Cramér-Rao bound were presented in [4, 8]. Still, the behavior of the specific algorithms is almost always investigated by simulation, either by the authors or in subsequent works [5].

Paper Organization: Section 2 describes the discrete network model we use. In Sec. 3 we analyze the expected uncertainty of the Intersection protocol. In Sec. 4 we analyze the expected uncertainty of the Arrangement protocol, comparing it with the results of Sec. 3. In Sec. 5 we describe the approximation protocol and show that its accuracy converges (with high probability) to that of the Arrangement protocol. We conclude the paper with Sec. 6, discussing some possible future research.

2   Discrete Network Model

Following [10], let $Q \in \mathbb{R}^2$ be a square subdivided into $(n+1)^2$ congruent squares, called *cells*, naturally indexed by the coordinates (i, j), where $-n/2 \leq i, j \leq n/2$ (w.l.o.g. we assume n is even).

We define the distance between any two cells $c = (i, j), c' = (i', j')$ to be $\mathrm{dist}(c, c') = \|c - c'\|_\infty = \max(|i - i'|, |j - j'|)$, and denote by R_{ij}^x the rectangle comprised of the set of cells that are at most x away from cell (i, j) (Fig. 1(a)),

$$R_{ij}^x = \{(i', j') | \max(|i - i'|, |j - j'|) \leq x\} \ .$$

When considering the set of cells that are at most x away from $(0,0)$ we omit the subscript, $R^x = R_{00}^x$. We denote the set of cells that are at least x away from the boundary by $Q_x = R^{n/2-x}$.

Generally we assume that N sensors, $S_1, \ldots, S_N$ are placed in Q, allowing several sensors to occupy the same cell. Two sensors can communicate with each other only if they occupy cells whose distance is less than or equal to the *communication range*, ρ, an integer between 0 and $n/4$. In this case we say that the two sensors are *within range* of each other and consider them to be *neighbors*.

Of the N sensors, only K know the coordinates of the cell they occupy (such sensors are called *anchors*). To avoid boundary conditions we assume that while the anchors are randomly placed in all of Q, the other $N - K$ sensors are randomly placed in the smaller region $Q_\rho{}^\star$. Furthermore, our expectation analysis is restricted to the sensors that are in $Q_{3\rho}$.

---

[*] This restriction is really relevant only for the Approximation protocol (Sec. 4), but it is easier to compare the protocols if we use the same boundary conditions all over.

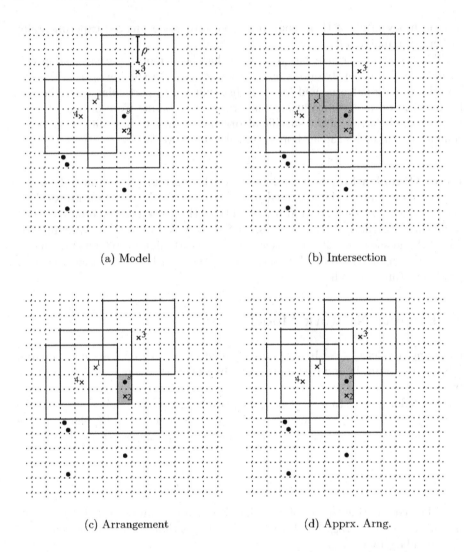

(a) Model   (b) Intersection

(c) Arrangement   (d) Apprx. Arng.

Fig. 1. Four anchors (numbered crosses) and several other sensors (bullets) are scattered in an $[n+1] \times [n+1]$ grid of congruent cells. Anchor x broadcasts its cell coordinates (i, j) in the rectangle, $R^\rho_{(i,j)}$, that is centered at (i, j) and has a side of size $2\rho + 1$ **(a)**. Sensor s can use the Intersection protocol **(b)** to position itself at the intersection of the rectangles related to anchors within range (1 and 2). It can use the Arrangement protocol **(c)** to position itself at the intersection of these rectangles that is not within range of any of the other anchors, 3 and 4. Or it can use the approximation **(d)** to locate itself in a slightly bigger region, defined in this case by the intersection of the ranges of anchors 1 and 2 that is not in the range of anchor 4 (anchor 3 is ignored because it is not a neighbor of the neighbors of s).

3 Localization by Intersection

The simplest localization approach for a sensor is to collect the positions of all its neighbor anchors and place itself somewhere at the intersection of the communication ranges of all these anchors; see Fig. 1(b). This is probably the most direct localization mechanism and it has been studied in [10]. Their analysis, however, is wrong since it neglected to take into account all the possible scenarios (for example, the case $m = 0$). In this section we introduce a new and simpler approach for the analysis of the expected size of this intersection, achieving correct results as opposed to [10].

Let S be a randomly picked sensor and denote by $S_{k_1}, \dots, S_{k_m}$ its anchor neighbors. Denoting by A_S the set of cells considered as possible locations for S by the protocol and by (x_i, y_i) the coordinates of S_{k_i},

$$A_S = \begin{cases} \bigcap_{i=1}^{m} R_{x_i y_i}^{\rho} & \text{if } m > 0, \\ Q_\rho & \text{Otherwise.} \end{cases}$$

That is, A_S includes all the cells at the intersection of the communication ranges of the neighbors of S if there are any, or the entire set of possible places in the grid otherwise.

Let $X = |A_S|$ be the number of different possible places for S. Then X is a random variable, $1 \leq X \leq (n - 2\rho + 1)^2$ and its expectation, $E(X)$ can be analyzed. As mentioned before, to avoid boundary affects we restrict the analysis to sensors that fall within $Q_{3\rho}$.

Let χ_{ij} be a family of indicator functions,

$$\chi_{ij} = \begin{cases} 1 & \text{cell } (i, j) \in A_S, \\ 0 & \text{Otherwise.} \end{cases}$$

Then

$$E(X) = \sum_{i=-n/2}^{n/2} \sum_{j=-n/2}^{n/2} E(\chi_{ij}) = \sum_{i=-n/2}^{n/2} \sum_{j=-n/2}^{n/2} \Pr[\chi_{ij} = 1] .$$

Cells that are outside $R_S^{2\rho}$ are clearly in A_S only if $m = 0$ (S has no neighbor anchors),

$$\Pr[\chi_{ij} = 1 | S \in Q_{3\rho}] = \left(1 - \frac{(2\rho+1)^2}{(n+1)^2}\right)^K, \quad (i, j) \in Q_\rho \setminus R_S^{2\rho} .$$

Translating the coordinates so S is at $(0, 0)$ we have

$$E(X | S \in Q_{3\rho}) = \left(1 - \frac{(2\rho+1)^2}{(n+1)^2}\right)^K \left((n - 2\rho + 1)^2 - (4\rho + 1)^2\right)$$

$$+ \sum_{i=-2\rho}^{2\rho} \sum_{j=-2\rho}^{2\rho} \Pr[\chi_{ij} = 1 | S \in Q_{3\rho}] .$$

Now if there is an anchor that is within range for cell $(0,0)$ but out of range for cell (i,j) then cell (i,j) is not considered a possible place for S. Hence, cell $(i,j) \in R_S^{2\rho}$ is in A_S only if all the K anchors fall outside $R_S^\rho \setminus R_{(i,j)}^\rho$; see Fig. 2(a). Because of the symmetry around the axes we can consider only the cases where $2\rho \leq i,j \leq 0$, in which case,

$$\Pr[\chi_{ij} = 1 | S \in Q_{3\rho}] = \left(1 - \frac{(2\rho+1)^2}{(n+1)^2} + \frac{(i+2\rho+1)(j+2\rho+1)}{(n+1)^2}\right)^K, \quad -2\rho \leq i,j \leq 0 .$$

Notice that if $i = j = 0$ then $R_S^\rho \setminus R_{(i,j)}^\rho = \emptyset$ so cell $(0,0)$ is always part of A_S. This fact is demonstrated by the final term in the expectation, which is 1.

$$E(X | S \in Q_{3\rho}) = \left(1 - \frac{(2\rho+1)^2}{(n+1)^2}\right)^K \left((n - 2\rho + 1)^2 - (4\rho + 1)^2\right)$$

$$+ 4 \sum_{i=-2\rho}^{0} \sum_{j=-2\rho}^{-1} \left(1 - \frac{(2\rho+1)^2}{(n+1)^2} + \frac{(i+2\rho+1)(j+2\rho+1)}{(n+1)^2}\right)^K + 1 .$$

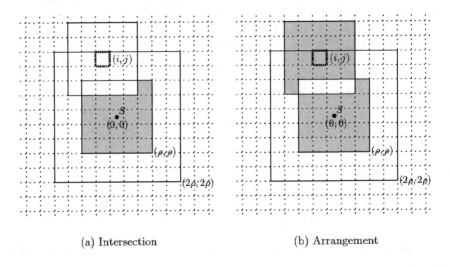

(a) Intersection   (b) Arrangement

Fig. 2. For the Intersection protocol cell (i,j), $-2\rho \leq i,j \leq 2\rho$ is in A_S if none of the K anchors fall in the shaded area in **(a)**. For the Arrangement protocol cell (i,j), $-2\rho \leq i,j \leq 2\rho$ is in A_S if none of the K anchors fall in the shaded area in **(b)**.

4   Localization by Arrangement

It is possible to improve the basic Intersection protocol if we assume that every sensor knows the position of all the K anchors (rather than just its neighbors).

This information can be gathered by flooding it over the network or by placing the sensors in several phases. With this knowledge a sensor can position itself at the intersection of the communication range of its anchor neighbors that is not within range of any other anchor; see Fig. 1(c). In this section we analyze the expected uncertainty of the Arrangement protocol and show that it achieves the same expected uncertainty as the Intersection protocol using at most half the anchors.

As before, let S be a randomly picked sensor, and denote by $S_{k_1}, \ldots, S_{k_m}$ its anchor neighbors and by $S_{k_{m+1}}, \ldots, S_{k_K}$ all the other anchors. Denoting by (x_i, y_i) the coordinates of anchor S_{k_i} and by A_S the set of possible places for S,

$$
A_S = \begin{cases} \bigcap_{i=1}^{m} R^{\rho}_{(x_i, y_i)} \setminus \bigcup_{i=m+1}^{K} R^{\rho}_{(x_i, y_i)} & \text{if } m > 0, \\ Q_\rho \setminus \bigcup_{i=m+1}^{K} R^{\rho}_{(x_i, y_i)} & \text{Otherwise .} \end{cases} \tag{1}
$$

That is, if S has neighbors then A_S consists of all the cells that are in range of the neighbor anchors of S and out of range of all the other anchors. If S has no neighbor anchors then A_S consists of all the possible cells in the grid that are outside the range of all the anchors.

Cells that are outside $R_S^{2\rho}$ can be in A_S only if $m = 0$ (S has no neighbor anchors) *and* they themselves do not have any anchor neighbors,

$$
\Pr[\chi_{ij} = 1 | S \in Q_{3\rho}] = \left(1 - \frac{2(2\rho+1)^2}{(n+1)^2}\right)^K, \quad (i,j) \in Q_\rho \setminus R_S^{2\rho} .
$$

Hence, assuming S is at $(0,0)$ and denoting as usual $X = |A_S|$,

$$
E(X | S \in Q_{3\rho}) = \left(1 - \frac{2(2\rho+1)^2}{(n+1)^2}\right)^K \left((n - 2\rho + 1)^2 - (4\rho + 1)^2\right)
$$

$$
+ \sum_{i=-2\rho}^{2\rho} \sum_{j=-2\rho}^{2\rho} \Pr[\chi_{ij} = 1 | S \in Q_{3\rho}] .
$$

Any cell $(i,j) \in R_S^{2\rho}$ is part of the area only if all the K anchors fall outside $(R_S^{\rho} \cup R^{\rho}_{(i,j)}) \setminus (R_S^{\rho} \cap R^{\rho}_{(i,j)})$; see Fig. 2(b),

$$
\Pr[\chi_{ij} = 1 | S \in Q_{3\rho}] = \left(1 - \frac{2(2\rho+1)^2}{(n+1)^2} + \frac{2(i+2\rho+1)(j+2\rho+1)}{(n+1)^2}\right)^K, \quad -2\rho \le i, j \le 0 .
$$

Using again the symmetry around the axes, we get that

$$
E(X | S \in Q_{3\rho}) = \left(1 - \frac{\mathbf{2}(2\rho+1)^2}{(n+1)^2}\right)^K \left((n - 2\rho + 1)^2 - (4\rho + 1)^2\right)
$$

$$
+ 4 \sum_{i=-2\rho}^{0} \sum_{j=-2\rho}^{-1} \left(1 - \frac{\mathbf{2}(2\rho+1)^2}{(n+1)^2} + \frac{\mathbf{2}(i+2\rho+1)(j+2\rho+1)}{(n+1)^2}\right)^K + 1 ,
$$

where the emphasized factors of 2 are the only difference in the expression from the expectation of the Intersection protocol.

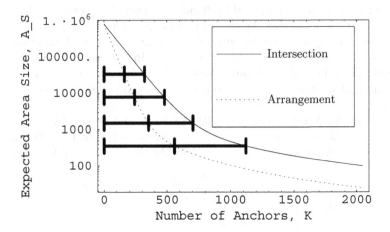

Fig. 3. A linear-log graph of the expected area size for the Intersection and Arrangement protocols ($n = 1000, \rho = 50$). The horizontal bars are aligned with the graph corresponding to the Intersection protocol and mark the middle point in term of K. As can be seen, the Arrangement protocol achieves the same expected accuracy with almost precisely $\frac{1}{2}K$ anchors.

4.1   Arrangement Versus Intersection

For any given ρ, n denote by E_I^K (E_A^K) the expectation of the area of uncertainty of the Intersection protocol (Arrangement protocol) using K anchors.

Let $c > 1$ be some constant s.t. $E_A^K \leq E_I^{cK}$. Denoting

$$x_{ij} = \frac{(2\rho+1)^2}{(n+1)^2} - \frac{(i+2\rho+1)(j+2\rho+1)}{(n+1)^2}, \quad -(2\rho+1) \leq i,j \leq 0$$

we require c to be such that for all i, j, $(1 - 2x_{ij})^K \leq (1 - x_{ij})^{cK}$. Noticing that $0 \leq x_{ij} \leq \frac{(2\rho+1)^2}{(n+1)^2} < 0.5$, it is not difficult to see that

$$(1 - 2x)^K \leq (1 - x)^{2K} \quad 0 < x < 0.5 \ ,$$

so that $c = 2$ will do. As can be seen in Fig. 3, the constant is almost precisely two for any reasonable choice of n, ρ.

5   Approximating the Arrangement

To consider a practical approximation for the arrangement protocol we assume that anchors can listen to the broadcasts of other anchors. Although at first anchors can only broadcast their own position, after an initial step every anchor

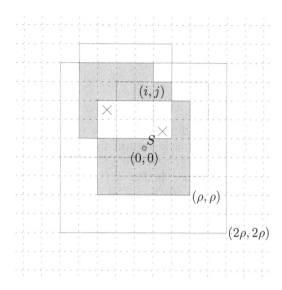

Fig. 4. Conditioned on the position of the two anchors in $R^\rho \cap R_{(i,j)}$ (marked by crosses), cell (i,j) is in A_S if the other $K-2$ anchors fall outside the shaded area.

can learn the position of its neighbor anchors and broadcast it along with its own position. A sensor can thus create two disjoint lists: a list of its neighbors and a list of its neighbor-neighbors that are not also its own neighbors. A sensor can now position itself at the intersection of its neighbors that is not within range of any of its neighbor-neighbors; see Fig. 1(d). As we show, the accuracy of this protocol converges to the accuracy of the arrangement protocol when the number of anchors increases.

Let S be a randomly picked sensor. Denote by $S_{k_1}, \ldots, S_{k_m}$ its anchor neighbors and by $S_{k_{m+1}}, \ldots, S_{k_{k+l}}$ all the other anchors that are also neighbors of at least one of $S_{k_1}, \ldots, S_{k_m}$. As before, denote by (x_i, y_i) the coordinates of anchor S_{k_i} and by A_S the set of possible places for S. Then

$$A_S = \begin{cases} \bigcap_{i=1}^{m} R^\rho_{(x_i,y_i)} \setminus \bigcup_{i=m+1}^{l} R^\rho_{(x_i,y_i)} & \text{if } m > 0, \\ Q_\rho & \text{Otherwise.} \end{cases} \tag{2}$$

That is, if S has neighbors then A_S consists of the cells that are within the range of the neighbors of S and outside the range of its neighbor neighbors. If S has no neighbors then A_S consists of all the possible cells in the grid.

When evaluating $E(|A_S|)$, one can use the same type of arguments as in Sec. 3 for cells that are outside $R_S^{2\rho}$ (such cells are in A_S only if $m = 0$). However, for cells $(i,j) \in R_S^{2\rho}$, the expectation is conditioned on the number and position of anchors in $R_{ij}^\rho \cap R_S^\rho$; see Fig. 4. It is cumbersome to analytically compute the expectation for this case. Instead, we show that for a large enough K the area of uncertainty returned by the Approximation protocol is equal with high probability to the area of uncertainty returned by the original Arrangement protocol.

5.1   Convergence

For any sensor S and any specific configuration of locations of anchors, denote by A_S the set of possible cells returned by the Arrangement protocol as defined in Eq. (1) and by A_S^{aprx} the set of possible cells returned by the Approximation protocol as defined in Eq. (2). Assuming as usual that S is at $(0,0)$ we require for simplicity that ρ is an integral multiple of 4. That is, there is an integer, r, s.t. $\rho = 4r$. Denote by I_1, I_2, I_3, I_4 the corner rectangles $R^r_{(-3r,-3r)}, R^r_{(3r,-3r)}, R^r_{(-3r,3r)}, R^r_{(3r,3r)}$ respectively; see Fig. 5.

Lemma 1. *If there is at least one anchor in each of I_1, I_2, I_3, I_4 then $A_S^{\mathrm{aprx}} = A_S$.*

Proof. Let $A_S^I = \bigcap\limits_{i=1}^{m} R^\rho_{(x_i,y_i)}$ be the intersection of the broadcast ranges of $S_{k_1}, \ldots, S_{k_m}$, the set of neighbor anchors of S. By definition every cell returned both by A_S and by A_S^{aprx} is in A_S^I. If some cell $c \in A_S^I$ is not in A_S^{aprx} then it must also not be in A_S, since it is in the broadcast range of anchors that are not neighbors of S. Hence $A_S^{\mathrm{aprx}} \supset A_S$.

Now let cell $c \in A_S^I$ be a cell such that $c \notin A_S$. Denoting by $A_S^U = \bigcup\limits_{i=m+1}^{K} R^\rho_{(x_i,y_i)}$ the union of the broadcast ranges of $S_{k_{m+1}}, \ldots, S_{k_K}$, all the other anchors, it follows that $c \in A_S^U$. Denote by S_* an anchor that is not a neighbor of S and whose broadcast range includes c. Since every cell in A_S^I must be at most ρ away from the anchors in I_1, I_2, I_3, I_4, cell c must be contained in the rectangle $R^{\rho/2}$, marked by the dashed inner rectangle in Fig. 5. But then anchor S_* must be in the annulus $R^{3/2\rho} \setminus R^\rho$, marked by the outer dashed rectangle in Fig. 5. Every cell in this annulus is within range of the anchors of at least one of I_1, I_2, I_3, I_4 so anchor S_* must be a neighbor of the neighbors of S which means that $c \notin A_S^{\mathrm{aprx}}$ so $A_S^{\mathrm{aprx}} \subset A_S$ and therefore the Lemma holds.

It is not hard to show that the probability that there is at least one anchor in each of I_1, I_2, I_3, I_4 is asymptotically 1. Hence, the output of the approximation protocol converges with high probability to that of the Arrangement protocol.

Theorem 1. *For any $\epsilon > 0$ there is some K_0 s.t. for all $K \geq K_0$, $A_S^{\mathrm{aprx}} = A_S$ with probability $1 - \epsilon$.*

Proof. The probability for any specific rectangle out of $I_1, \ldots, I_4$ to remain empty is $(1 - \frac{(2r+1)^2}{(n+1)^2})^K$. Denoting by X the number of such empty rectangles, it follows that $\mathrm{E}(X) = 4(1 - \frac{(2r+1)^2}{(n+1)^2})^K$ so, using the Markov bound,

$$\Pr[X \geq 1] \leq \mathrm{E}(X) = 4(1 - \frac{(2r+1)^2}{(n+1)^2})^K \leq 4e^{-\frac{(2r+1)^2}{(n+1)^2}K}.$$

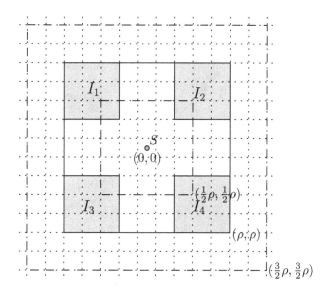

Fig. 5. If there is at least one anchor in each of the rectangles I_1, I_2, I_3, I_4 then the set of possible positions for S returned by the Approximation protocol is equal to that returned by the Arrangement protocol.

Requiring that $\Pr[X \geq 1] \leq \epsilon$ we get $K_0 = \left(\frac{n+1}{\rho/2+1}\right)^2 \ln \frac{4}{\epsilon}$ and hence the Theorem holds.

In practice $E(A_S^{\mathrm{aprx}})$ converges quite rapidly. Figure 6 presents simulation results for the expected uncertainty of the Approximation protocol plotted against the expectations of the Intersection and Arrangement protocols. As can been seen, the simulated expectation converges rapidly to that of the Arrangement protocol. It should be noted that when K is small, all three expectations are dominated by the event where S has no neighbor anchors ($m = 0$). In this case the size of the area of uncertainty is the entire grid, which is typically very large compared to the other cases. As more anchors are placed, the probability of this event quickly diminishes to the point where it has negligible effect. This behavior, however, creates an anomaly in the shape of a phase transition in a straightforward simulation. Therefore, the data presented in Fig. 6 is the simulation results conditioned on S having neighbors analytically adjusted to take into account the case where $m = 0$.

6   Conclusions

We consider three range-free localization protocols for sensor networks and analyze their expected positioning accuracy under random sensor distribution. Assuming a small set of anchor nodes that know their position and broadcast it, we

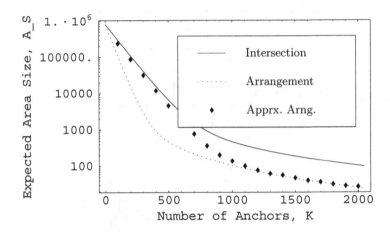

Fig. 6. Simulation results for the expected uncertainty of the Approximation algorithm are plotted against the expectations of the Intersection and Arrangement protocols ($n = 1000, \rho = 50$). The simulated expectation converges rapidly to that of the Arrangement protocol.

investigate at first the case where sensors position themselves at the intersection of the broadcast range of the anchors they detect (their neighbors). We then consider the positioning uncertainty when sensors avoid positioning themselves in places that are in the range of non-neighbor anchors, and an approximation for the second case where the proximity of only some local subset (the neighbor neighbors') of all the non-neighbor anchors is avoided.

An interesting extension to this work would be to assume that the broadcast range is defined by a disk (l^2-norm) in the continues model. We plan to investigate this case in the future, looking in particular at the connection to stochastic geometry.

References

1. N. Bulusu, V. Bychkovskiy, D. Estrin, and J. Heidemann. Scalable, ad hoc deployable rf-based localization. In *Grace Hopper Celebration of Women in Computing Conference 2002*, Vancouver, British Columbia, Canada, October 2002.
2. N. Bulusu, J. Heidemann, and D. Estrin. Gps-less low cost outdoor localization for very small devices. *IEEE Personal Communications Magazine*, 7(5): 28–34, October 2000.
3. L. Doherty, L. E. Ghaoui, and S. J. Pister. Convex position estimation in wireless sensor networks. In *IEEE Infocom*, volume 3, pages 1655–1663, April 2001.
4. Hanbiao Wang, Len Yip, Kung Yao, and Deborah Estrain. Lower bounds of localization uncertainty in sensor networks. In *IEEE ICASSP*, May 2004.

5. Koen Langendoen and Niels Reijers. Distributed localization in wireless sensor networks: a quantitative comparison. *Elsevier Computer Networks*, 43(4):499–518, 15 Nov. 2003.

6. R. Nagpal, H. Shrobe, and J. Bachrach. Organizing a global coordinate system from local information on an ad hoc sensor network. In *Information Processing in Sensor Networks: Second International Workshop*, IPSN 2003, number 2634 in Lecture Notes in Computer Science, pages 333–348, Palo Alto, April 2003. Springer-Verlag.

7. D. Niculescu and B. Nath. Dv based positioing in ad hoc networks. *Telecommunication Systems*, 22(1–4):267–280, 2003.

8. N. Patwari, A. III, M. Perkins, N. Correal, and R. O'Dea. Relative location estimation in wireless sensor networks. *IEEE Transactions on Signal Processing*, 51(8):2137–2148, August 2003.

9. Y. Shang, W. Ruml, Y. Zhang, and M. P. J. Fromherz. Localization from mere connectivity. In *Proceedings of the fourth ACM international symposium on Mobile ad hoc networking & computing, MOBIHOC 2003*, pages 201–212. ACM Press, 2003.

10. S. Simic and S. Sastry. Distributed localization in wireless ad hoc networks. Technical Report UCB/ERL M02/26, UC Berkeley, 2002.

Towards a Dynamical Model for Wireless Sensor Networks

Pierre Leone and José Rolim

Computer Science Department, University of Geneva
24 rue Général Dufour 1211 Geneva 4
Switzerland {pierre.leone,jose.rolim}@cui.unige.ch

Abstract. In this paper we introduce a dynamical model for wireless sensor networks. We obtain a convergent martingale for the broadcast process in such networks. To our knowledge, such martingales were unknown previously. We look at a formal model using the formalisms of martingales, dynamical systems and Markov chains, each formalism providing complementary and coherent information with each other. The model is partly validated with numerical simulation of wireless sensor networks, we informally make explicit the situations where the model is realistic. We also provide an alternative dynamical model based on the hypothesis that the distribution of the location of the emitting sensors is uniform. This hypothesis is more fulfilled when the emission radius r becomes larger and emission angle α smaller. In the situations where the hypothesis is close to be fulfilled good agreements between numerical epxeriments and results of the model are observed. The alternative approach leads to similar quantitative and qualitative results as the first model.

1 Introduction

In this paper we propose a model for describing dynamical aspects of broadcast in wireless sensor networks. To our knowledge works on modeling wireless sensor networks focus on static aspects, for instance the topology given by the underlying graph of connections[6], or focus on specific algorithms such as routing [5, 8], localization [4], etc.

The network we consider is a set of sensors equipped with directional antenna allowing the transmission of data in a sector of disk described by radius r and emission angle α which are the same for all sensors. The direction of the emission is characterised by angle β which is a random variable uniformly distributed on $[0, 2\pi[$ and fixed for all emissions (each sensor possesses its own direction of emission), see Figure 1. Note that when $\alpha = 360$ we get a classical sensor network without directional antenna [2]. Thus, our model is general.

The transmission of data is through radio waves, the frequency of emission is the same for all sensors and so collisions or interferences have to be taken into account. Collisions occur when at least two sensors send data to the same third

S. Nikoletseas and J. Rolim (Eds.): ALGOSENSORS 2004, LNCS 3121, pp. 98–108, 2004.

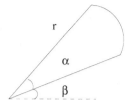

Fig. 1. Emission radius r and angle α with direction angle β.

sensor. In this case, the receiving sensor is able to detect that the data is unco-
herent. However, emitting sensors are unable to detect that collision occurred. In
Figure 2, x and y are emitting sensors, each sensor located in the hatched region
cannot receive the data because of the collisions. We point out that the process
of interference of electromagnetic waves is linear, subject to the superposition
principle, and so the hatched region is the only one in which collision occurs.

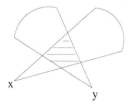

Fig. 2. x, y are the emitting sensors and the hatched region represents the collision
region.

The region of space in which is deployed the network is supposed to be a square
region of unit surface, sensors are thrown away randomly with a uniform distribution (for a complementary presentation of the network see [1]). Our first
model ignores border effects due to sensors located closely to the border of the
region. For this, we consider a square region where we identify opposites boundaries, actually removing the borders. This is equivalent to look at the network
deployed on a sphere.

The broadcast we consider is the simplest one: A sensor desires to broadcast
data through the entire network. At time zero data is emitted. At the next step,
all sensors which just received the data send it in turn and so on. Emissions are
assumed to occur at discrete time $n = 0, 1, 2, \ldots$ and reception occur during a
given interval of time say $[n, n+1[$. We look at our model under different points of
view, using martingales, dynamical systems and Markov chain formalism. Each
approach shed light on the behaviour of the model consistently with each others.
The underlying graph we get through the process of throwing away sensors and
connecting sensors which can send/receive data (not necessarily the directed

graph) is called random scaled sector graph and leads to important work, cf [6]. Understanding the underlying graph is certainly of prime importance and subsequent models would include such information.

2   Broadcast and Martingales

We consider a network of N sensors, each sensor with emission radius r, emission angle α and direction angle β (β different for each sensor). Let us note by $p = \frac{1}{2}\alpha r^2$ the area covered by an emitting sensor. Due to the uniform distribution of sensors on the whole unit area, the number of sensors in the area is a binomial random variable denoted by $B(N, p)$. So, the probability for a sensor to belong to the area covered by another sensor is p and the probability that it does not belong $q = 1 - p$. Therefore, p is the probability that a sensor receives data from an emitting sensor. We now formulate the main hypothesis of our model, namely we assume the probability of the event of belonging to the area covered by a given sensor is independent of the probability to belong to the area covered by another one. This hypothesis leads to our underlying graph to be a classical random graph [3] (probability p of an edge between two sensors). Let us now introduce the random variable Z_n counting the number of emitting sensors at time n and X_i^n, $i = 1, \ldots, Z_n$ count the number of sensors receiving data from the i-th emitting sensor. Because of possible collisions, receiving data from the i-th sensor means not be covered by another emitting one. Because there are Z_n emitting sensors the probability to be covered by one and only one emitting sensor is pq^{Z_n-1} (with our independence hypothesis) and so the X_i^n's are distributed following a $B(N, pq^{Z_n-1})$. We are now able to make explicit the distribution law of Z_{n+1} given Z_n. Indeed, Z_{n+1} is given by

$$Z_{n+1} = \sum_{i=1}^{Z_n} X_i^n, \qquad (1)$$

and so its distribution is $B(NZ_n, pq^{Z_n-1})$. Next proposition shows that $Z_n \leq N$ with high probability (w.h.p.) which is a necessary condition for our model to be meaningful. The result depends on the assumption that $p < .82$ for technical reasons. Intuitively this can be understood considering the expectation of Z_{n+1}, $E(Z_{n+1}) = NZ_n pq^{Z_n-1}$. In order to fulfill $Z_{n+1} < N$ w.h.p we have certainly to fulfill the condition $Z_n pq^{Z_n-1} < 1$ and this leads to the constraint on p. We point out that this assumption is not stringent as easily satisfied in practice where p is usually less than .1. Indeed, when working with wireless sensor networks we generally assume that p is small with respect to the total unit area [2].

Proposition 1. *We consider the stochastic process* $\{Z_n\}_{n \geq 0}$ *where* Z_{n+1} *is distributed following a* $B(NZ_n, pq^{Z_n-1})$ *where we assume* $p < .82$, $q = 1 - p$. *Then,*

$$P(Z_{n+1} > N) \leq e^{-N(\gamma - \ln(\gamma) - 1)}, \qquad (2)$$

with $\gamma \leq -\frac{p}{\log(1-p)(1-p)e}$ *and* $\gamma - \ln(\gamma) - 1 > 0$.

Proof. We know from [7] that

$$P(Z_{n+1} > N) \le e^{-cN} \prod_{i=1}^{Z_n N} \left(1 + pq^{Z_n-1}(e^c - 1)\right), \tag{3}$$

with $c > 0$. So, using $1 + x \le e^x$ we get

$$P(Z_{n+1} > N) \le e^{-N(c - Z_n pq^{Z_n-1}(e^c-1))} = e^{-N(c-\gamma(e^c-1))}, \tag{4}$$

with $\gamma = Z_n pq^{Z_n-1}$ which is maximal for $Z_n = -\frac{1}{\log(q)} = -\log_q(e)$ and so $\gamma \le -\frac{p}{\log(q)qe}$. Provided $p < .82$ we have $\gamma < 1$. The term $c - \gamma(e^c - 1)$ is maximal for $c = -\log(\gamma) > 0$, then introducing this particular value for c leads to the result.

To set up a martingale framework, let us introduce the σ-algebra generated by the random variables $Z_0, Z_1, \ldots, Z_n$ denoted by $\mathcal{F}_n = \sigma(Z_0, Z_1, \ldots, Z_n)$. We then get a filtration $\mathcal{F}_0 \subset \mathcal{F}_1 \subset \mathcal{F}_2 \subset \ldots$ and computing conditional expectation of Z_{n+1} with respect to $\mathcal{F}_n$, we get

$$E(Z_{n+1} \mid \mathcal{F}_n) = \frac{Np}{q} Z_n q^{Z_n}. \tag{5}$$

Taking this into account we can prove by direct computation that M_{n+1} defined as

$$M_{n+1} = \frac{Z_{n+1}}{(\frac{Np}{q})^n q^{Z_0+\ldots+Z_n}}, \tag{6}$$

is a martingale with respect to the defined filtration, i.e.

$$E(M_{n+1} \mid \mathcal{F}_n) = M_n. \tag{7}$$

Moreover, because the martingale is positive then it is convergent [10]. We summarise all these results in the next proposition.

Proposition 2. *Consider the stochastic process $\{Z_n\}_{n\ge 1}$ with $Z_0 = 1$ and Z_{n+1} is distributed following a $B(NZ_n, pq^{Z_n-1})$. Then,*

$$M_{n+1} = \frac{Z_{n+1}}{(\frac{Np}{q})^n q^{Z_0+\ldots+Z_n}}, \tag{8}$$

is a convergent martingale with respect to the filtration generated by $Z_0, Z_1, \ldots, Z_n$. Convergent means that $\lim_{n\to\infty} M_n = M_\infty$ almost surely.

Proof. The fact that M_{n+1} is a martingale follows by direct computations. The martingale is almost surely convergent because of positivity [10].

3   Broadcast and Dynamical Systems

The convergent martingale introduced in the last section gives us some insight into the dynamic of the broadcast process. Because of the almost sure convergence $M_n \to M_\infty$ we write

$$M_n \approx c, \tag{9}$$

with n large enough and c a constant. We assume that $c \neq 0$ and look at the quotient $\frac{M_{n+1}}{M_n}$. This leads to

$$Z_{n+1} = (\frac{Np}{q})Z_n q^{Z_n} = f(Z_n). \tag{10}$$

We can drop the mentioned hypothesis $c \neq 0$ arguing that the difference $M_{n+1} - M_n$ tends to vanish with increasing n because of the almost sure convergence of the martingale. Direct computations lead again to equation (10).

Equation (10) suggests that the behaviour of the stochastic process $\{Z_n\}$ is related to the dynamical system $Z_n = f^n(Z_0)$ with f^n referring to the n-th composition of f. Note that $f(Z_n) = E(Z_{n+1} \mid \mathcal{F}_n)$ which is nothing else than the best estimator, with respect to the L^2 measure, of Z_{n+1} at time n. Actually, our model can be seen as a random perturbation (see [9]) of the deterministic dynamical system defined by (10). In the following, we take into account only the deterministic behaviour and we validate with numerical experiments in the next section that the qualitative behaviour seems not to be changed with the random perturbation. Although quite informal this approach gives an important insight into the behaviour of our model.

As classical in the literature on discrete dynamical systems, looking at iterations of the function $f^n(Z_0)$ can be represented graphically with the following procedure. Let us start from the point Z_0, computing $f(Z_0)$ amounts to go vertically from the 'x-axis' at position $x = Z_0$ to the curve $y = f(x)$. Next, we move horizontally up to crossing the line $y = x$, this amounts to project the y-axis on the x-axis. Then we continue moving vertically and so on. In Figure 3, this process is depicted.

We observe that the curves $f(x)$ and $y = x$ intersect at $x = 0$ and $x = \log_q(\frac{q}{Np})$ which are fixed points. To refer to any fixed points we note x_0. The behaviour close to the fixed points x_0 is determined by $f'(x_0)$. The graphical process sketched above should be enough to convince the reader that if $\mid f'(x_0) \mid < 1$ then x_0 is attractive and if $f'(x_0) > 1$, x_0 is repulsive.

Let us now consider $x_0 = 0$, direct computation shows that $f'(0) = \frac{Np}{q}$. In order to ensure broadcast the fixed point $x_0 = 0$ must be repulsive and so parameters N and p must satisfy

$$\frac{Np}{q} > 1, \tag{11}$$

which is a first constraint on our network. Actually, this condition is quite natural because Np is the expected number of sensors in the area of the first emitting sensor.

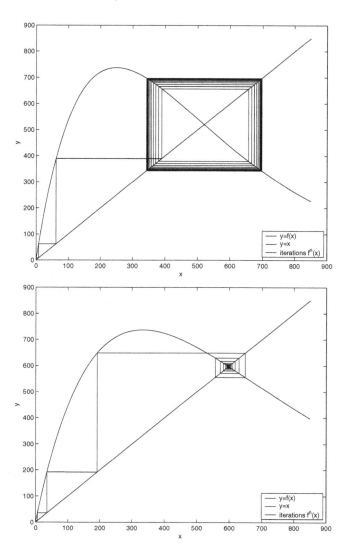

Fig. 3. Graph of $f(x)$ with iterations, $p = 0.004$ on the top and $p = 0.003$ above. In both cases N=2000.

Let us now look at the second fixed point $x_0 = \log_q(\frac{q}{Np})$. At x_0 we have $f'(x_0) = 1 + \log(\frac{q}{Np})$ and because of our previous condition (11), we have $f'(x_0) < 1$. Direct computations show that if $p < \frac{e^2}{N+e^2}$ then x_0 is attractive. In that situation, iterations of $f(x)$ tend to the fixed point. This is illustrated on the bottom of Figure 3 ($p = 0.003$). On the top of Figure 3 we have $p = 0.004$ and then $f'(x_0) < -1$ implying that x_0 is repulsive. In this situation we observe that a cycle appears, the behaviour of the dynamical system tends to be periodic with period two ($f^{n+2}(x) \approx f^n(x)$). It is important to notice that the appearance of a cycle depends crucially on the graph of $f(x)$ in a non trivial way. We cannot assert many results on the behaviour of the dynamical system provided $f'(x_0) < -1$ especially when $f'(x_0)$ is very far from -1. However, we can assert a necessary condition for broadcast. Function $f(x)$ is maximal for $x_m = -\frac{1}{\log(q)} = -\log_q(e)$ and it is clear that $f(x_m) < 1$ implies that the broadcast is uncertain to go on because when the iteration of the function come close to x_m, the next step the predicted number of emitting sensor will be smaller than one and therefore the broadcast process ends.

4   Broadcast and Markov Chain

In our previous analysis we looked at deterministic aspects of the dynamic letting aside the question of the qualitative behaviour under stochastic perturbations. In order to investigate how stochastic perturbations act on our model we find useful to reformulate our model as a Markov chain and numerically investigate the invariant measure.

We still consider $\{Z_n\}_{n \geq 0}$ with Z_{n+1} distributed as $B(NZ_n, pq^{Z_n-1})$. Then, the transition matrix $P = (p_{ij})$ with

$$p_{ij} = P(Z_{n+1} = j \mid Z_n = i) = (N)\,ij\,(pq^i)^j\,(1 - pq^i)^{Ni-j}. \qquad (12)$$

We numerically look for an invariant measure for the Markov chain computing the powers of P with parameters $N = 2000$, $r = 0.1$ and $\alpha = .8$ where N is the total number of sensors, r the radius of the emission and α the angle of the emission. This amounts to choose $p = 0.04$. From our previous analysis we expect to detect the appearance of a cycle of periodicity two (see Figure 3, left side). Result is depicted in the left side of Figure 4 and shows a bimodal curve which is in accordance with what is expected from our previous analysis. In the right side of Figure 4 is depicted the invariant measure with the same parameters as before but with $\alpha = 0.6$ ($p = 0.003$). We observe a unimodal curve centered around $Z \approx 596$ which is very close to the value of the fixed point (≈ 597) predicted with the dynamical approach of the previous section. This means that the random perturbation does not change the qualitative behaviour of the deterministic model.

5   Numerical Experiments

In this section we report on the numerical experiment we have made concerning the simulation of the broadcast process. The program throw out randomly (uniform distribution on a square on unit surface) N sensors together with the angles of emission (uniform distribution on $[0, 2\pi[$). Then, we select randomly one sensor which initiates the broadcast process. At each time we record the number of emitting sensors and the result is displayed as an histogram on Figure 5.

We observe that the simulation for $p = 0.003$ is very similar to the prediction of the model, see Figure 4 supporting the model for small values of p.

However, for the larger value of $p = 0.004$ where the model predicts oscillation and bimodal invariant curve, see Figure 3, numerical simulation shows a unimodal curve. The qualitative behaviour observed on numerical experiments does not mimic what is expected from the analysis of the model.

Actually, it happens that the behaviour of the broadcast process depends on both parameters r and α and not only on the product $p = \frac{1}{2}\alpha r^2$. In the next section we provide simulations accurately described by the model and explain in which situations the model is more accurate. The discussion leads to tracks for our further work.

Nevertheless, we point out that in any cases the mean value which is predicted with the model (≈ 525, see Figure 4) is very close to what is expected from numerical experiments (compare Figure 4 with Figure 5).

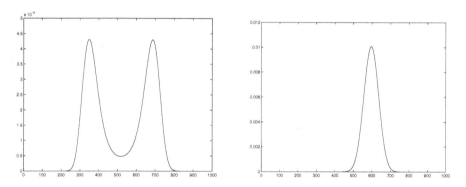

Fig. 4. Invariant measure for the Markov chain (12) with $p = 0.004$ (left) and $p = 0.003$ (right).

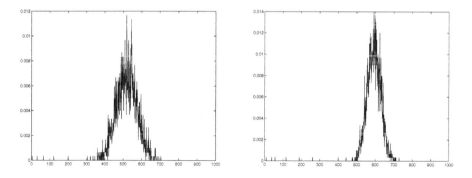

Fig. 5. Simulation of the broadcast process with $N = 2000$, $r = 0.1$ and $\alpha = 0.8$ (left) $\alpha = 0.6$ (right)

6   Alternative Approach

In this section we introduce different assumptions to look for a model of the broadcast process as the ones introduced in section 2. The derived model is slightly different than the one described up to now in this paper but qualitatively and quantitatively very similar. Results are in the same spirit as the rest of the paper and details are omitted. This is of great help in order to determine the key factors leading to the behaviour described in the previous sections and enables us to understand conditions on the emitting radius and angle r, α, such that the model is accurate.

Let us denote Z_n the number of emitting sensors at time n. Our main hypothesis is now that the emitting sensors at time n are uniformly distributed. Intuitively, this assumption is satisfied when r is large with respect to the dimension of the whole area considered and α small. This means that a more realistic model should take into account values of r and α independently and not only of the product $\frac{1}{2}\alpha r^2$. Given a fixed value of $p = \frac{1}{2}\alpha r^2$, our model is more accurate as r gets larger values and α smaller ones keeping p constant.

At time n, let us consider sensor x_i. The number of emitting sensors Y_i^n able to transmit data to x_i is then a random variable $Y_i^n \approx B(Z_n, \pi r^2)$. We introduce X_n^i the random variable which is one if one and only one emitting sensor send data to x_i, we know that X_i^n is a Bernouilli random variable with probability $Y_i^n \frac{\alpha}{2\pi}(1 - \frac{\alpha}{2\pi})^{Y_i^n-1}$ of being one. Conditioning on Y_i^n we can compute the expectation of X_i^n,

$$E(X_i^n \mid Z_n) = \frac{\alpha r^2}{2}Z_n(1 - \frac{\alpha r^2}{2})^{Z_n-1} = Z_n p q^{Z_n-1}, \tag{13}$$

with $p = \frac{\alpha r^2}{2}$, $q = 1 - p$ as usual in this paper, see section 2. The number of emitting sensors at time $n+1$ is then a random variable given by (compare with (1))

$$Z_{n+1} = \sum_{i=1}^{N} X_i^n. \tag{14}$$

So the distribution of Z_{n+1} is a binomial $B(N, Z_n p q^{Z_n-1})$. We can now apply the same procedure as the one described in section 2 to show that the same martingale M_n has to be considered here again.

The Markov chain formalism is slightly different than the one introduced in section 4, see formula (12), in our context we get

$$p_{ij} = P(Z_{n+1} = j \mid Z_n = i) = (N) j(ipq^{i-1})^j(1 - ipq^{i-1})^{N-j}. \tag{15}$$

However, numerical investigation of the invariant measure in the same conditions ($r = 0.1$ amd $\alpha = 0.6, 0.8$) leads to very similar curves than the ones depicted in Figure 4.

Actually, the fact that the martingale is the same for both models implies at once than the behaviour of the fixed points are the same when we consider the deterministic dynamical system. However, it was not evident that the random perturbation (invariant measure) are similar in both cases.

In the the light of this section we consider again numerical simulations of the broadcast process. We still consider $p = 0.004$ in order to compare with previous numerical experiments but we now choose $r = 1$ large and $\alpha = 0.008$ small. In this situation we expect the model to be accurate because we are in the situation where the emitting sensors tend to be uniformly distributed. Numerical experiments are depicted in Figure 6 and shows good agreements between simulations and model (compare with Figure 4). These experiments show that a realistic model should depends on the parameters r and α independently and not only on the product $\frac{1}{2}\alpha r^2$.

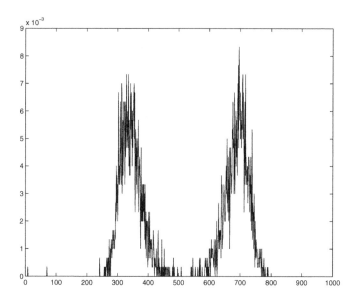

Fig. 6. Simulation of the broadcast process with $N = 2000$, $r = 1.0$ and $\alpha = 0.008$.

7   Conclusion and Future Work

In this paper we formulate a first dynamical model for broadcast in wireless sensors network. The model is simple enough to be analyzed numerically, and shows interesting qualitative behaviour. For small values of p where the model does not predict oscillations in the number of emitting sensors we observe a good agreement of the model with numerical simulations, compare Figure 4 with Figure 5 right side. In general, as p increases, oscillations predicted with the model are not observed with numerical simulations. This shows that the broadcast process depends independently on the parameters r and α and not only on the product $p = \frac{1}{2}\alpha r^2$ as the model. However, we show that the model is accurate for r large with respect to the dimension of the considered whole area and α small. More precisely, given a fixed value of $p = \frac{1}{2}\alpha r^2$ the model is more accurate as r gets larger values and α smaller ones in order to keep p constant. This is the result of the alternative approach we formulate in section 6 based on the uniform distribution of the emitting sensors. In this situation, oscillations are numerically observed and both qualitatively and quantitatively well described with our model, compare Figure 4 with Figure 6. Further work will include independent dependencies of the model on the parameters r and α.

References

1. G. Arzhantseva, J. Diaz, J. Petit, J. Rolim, M.J. Serna. Broadcasting on Networks of Sensors Communicating through Directional Antennas. Crescco Technical Report, 2003. available on the web at http://www.ceid.upatras.gr/crescco/
2. A. Bharathidasan, Vijay Anand Sai Ponduru, Sensor Networks : An Overview, available on the web at http://wwwcsif.cs.ucdavis.edu/ bharathi/sensor/snw.html.
3. B. Bollobas, Random Graphs. Academic Press, London, 1985.
4. N. Bulusu, J. Heidmann, D. Estrin, GPS-less Low Cost Outdoor Localization For Very Small Devices, IEEE Personal Communications, Special Issue on Smart Spaces and Environments, Vol. 7, No. 5, pp. 28-34, October 2000.
5. D. Braginsky, D. Estrin, Rumor Routing Algorithm for Sensor Networks, In Proceedings of the First Workshop on Sensor Networks and Applications (WSNA), September 28 2002, Atlanta, GA.
6. J. Díaz, V. Sanwalani, M. Serna, P. Spirakis. Chromatic number in scaled sector graphs, available on the web at http://www.lsi.upc.es/ diaz/papers.html.
7. J. Díaz, J. Petit, M. Serna. A guide to concentration bounds. In S. Rajasekaran, P. Pardalos, J. Reif, and J. Rolim, editors, Handbook on Randomized Computing, volume II, chapter 12, pages 457-507. Kluwer Press, New York, 2001.
8. C. Intanagonwiwat, R. Govindan, D. Estrin, Directed Diffusion : A Scalable and Robust Communication Paradigm for Sensor Networks, In Proceedings of the Sixth Annual International Conference on Mobile Computing and Networks (MobiCOM 2000), August 2000, Boston Massachusetts.
9. Y. Kiefer. Random Perturbations of Dynamical Systems, Birkhäuser Boston, 1988. Handbook of
10. D. Williams, Probability with Martingales . Cambridge University Press, 1991.

Efficient Delivery of Information in Sensor Networks Using Smart Antennas

Tassos Dimitriou and Antonis Kalis

Athens Information Technology
Markopoulo Ave., 19002, Athens, Greece
{tdim, akal}@ait.gr
http://www.ait.gr

Abstract. In this work we present a new routing protocol for sensor networks that utilizes smart antennas to propagate information about a sensed event towards a receiving center. Our protocol is suited for those cases where unexpected changes to the environment must be propagated quickly back to the base station without the use of complicated protocols that may deplete the network from its resources. The novelty of our approach lies in the fact that our protocol uses only local information and total absence of coordination between sensors; during a simple initialization phase each node uses the beam that lies towards the direction of the base station to transmit data and the beam lying on the opposite side of the plane to receive data. We provide detailed experimental analysis that demonstrates the feasibility of this approach, the necessity of using smart antennas in sensor networks and the advantages that are presented to communications due to their use. In particular, we demonstrate that sensed data are propagated by activating only the sensors that lie very close to the optimal path between the source of the event and the destination, resulting in low activation of the network's sensors. Furthermore, our protocol is very easy to implement and more importantly it is scalable as it remains independent of network size.

1 Introduction

Sensor networks [1,2] have attracted much scientific interest during the past few years. Networks of thousands of sensors may represent an economical solution to some challenging problems: real-time traffic monitoring, building safety monitoring, wildlife monitoring, fire sensing, movement tracking, etc. These networks differ from wireless ad hoc networks in that their nodes are more energy constrained; nodes employed in sensor networks are characterized by limited resources such as storage, computational and communication capabilities. The power of sensor networks, however, lies exactly in the fact that their nodes are so small and cheap to build that a large number of them can be used to cover an extended geographical area, gather information in-site and process it in parallel enabling an accurate and reliable monitoring process. And is exactly this data delivery aspect that is the most common characteristic of sensor networks.

S. Nikoletseas and J. Rolim (Eds.): ALGOSENSORS 2004, LNCS 3121, pp. 109–122, 2004.

Data, dynamically acquired from the environment, travel through the network towards some base station, offering low-latency real-time information that was previously hard or infeasible to get.

There are basically three types of schemes [3] concerning data delivery: continuous, event driven and observer-initiated. According to the first one, sensor nodes send their measurement to the base station at a specified rate, while in the event-driven model nodes send the measured data to the base station whenever they detect some type of activity that is worth reporting. In the observer-initiated scheme, the base station itself issues queries to any node in the network or to all nodes within a specific area, resulting in sensors collecting data and sending them back to the base station. Due to the limited resources available to nodes however, expensive routing protocols, costly flooding mechanisms, or complex algorithms that don't scale to large number of nodes cannot be used. Furthermore, random distribution of nodes in the physical environment, node failure probability during their deployment and dynamic change of nodes' power supply make the design of communication protocols a very challenging task.

Here we focus on the efficient propagation of a sensed event towards some receiving center, assuming an event-driven data delivery model. The need for communication between a regular sensor (the *source*) and some base station (called the *destination* or the *sink*) can arise at any time, possibly triggered by unexpected changes in the environment. It is exactly this change in the environment (a fire, a person entering a restricted area, etc.) that we feel it is important to reach the base station as quickly as possible without of course depleting the network from its resources through the use of complicated protocols. So, our focus is the design of localized algorithms where nodes collaborating with their neighbors achieve a *global objective*, that of delivering measured data to the base station.

In this paper we consider the use of smart antenna systems in order to achieve reliable and efficient data delivery in wireless sensor networks. Smart antennas in general have been for long considered unsuitable for integration in wireless sensor nodes. They consist of more than one antenna element and therefore require a larger amount of space than traditional antennas. In addition to that, processing of more than one signal requires more computational power and electronics that translate radio frequency signals to baseband signals suitable for processing. In this paper, however, we show that the use of smart antennas in sensor networks is in some cases obligatory and in other cases achievable, with minimal additional cost.

The rest of the paper is organized as follows. In Section 2 we examine the ability of sensor nodes to integrate smart antenna systems and the benefits of using such schemes in wireless ad-hoc communications. Is Section 3 we present the proposed routing algorithm that utilizes smart antenna systems, and in Section 4 we present the benefits of using this algorithm in the network layer of the sensor network. Finally we conclude our work and present directions for future research in Section 5.

2   Applying Smart Antennas in Sensor Networks

The aim for the future is to create sensor nodes (a.k.a. smart dust) that measure no more than 1 cubic millimeter in volume [4]. Sensor node architectures are currently advancing towards meeting these requirements [5]. In order to achieve efficient wireless communication in smart dust dimensions, one of two wireless mediums must be used: optics or radio frequency (RF).

Optical wireless is a promising technology for wireless communications. The dimensions of light emitting diodes (LEDs) and laser-diodes have reached sub-millimeter levels, and micro electromechanical systems (MEMS) technology has enabled these devices to be implemented with very low cost. Optical wireless is considered an optimal solution for certain sensor network environments. In in-flight sensor nodes, for example, communication is achieved using LEDs or laser beams that have large directivity even when diffused infrared (DI) methods are used. Using on-off-keyed signals to achieve omni-directional (OD) coverage, multiple infrared sensors and actuators are deployed inside sensor nodes. Steering and switching among directed beams is performed to find the best beam possible. In essence, these optical wireless transceivers are switched beam smart antennas that ensure OD coverage. Unfortunately, the main drawback of these optical wireless technologies is that optical communications are greatly affected by the environment. Oftentimes ambient lighting conditions produce interference and connections are cut abruptly when line of sight (LOS) signals are disrupted.

Furthermore, the use of narrow fixed beams in nodes that are randomly distributed in a certain geographical area does not provide the necessary connectivity to achieve data propagation through the network [6]. For example, after extensive simulations we have seen that using a simple flooding mechanism and directional beams with 60 degrees width the packet delivery ratio (PDR) does not surpass 40%, while the use of omni-directional antennas achieves 100% probability of delivery.

Radio frequencies are widely used for wireless communications in current sensor networks. RF presents several advantages with respect to other transmission techniques in terms of range, coverage, and power efficiency. Current implementations of sensor nodes use a single OD antenna to achieve coverage. It can be argued that smart antennas are not suitable for sensor nodes since they consist of more than one element and are therefore larger. However, when we refer to sensor nodes, we refer to three-dimensional structures. If a single- or two-dimensional half or quarter of a wavelength antenna can fit in a sensor node then more than one of these antennas should be able to fit into a three-dimensional node. In addition, inter-element spacing of a half or quarter wavelength between antennas should be possible (details omitted here). Since more than one antenna can fit in the same three-dimensional space, provision for patterns with larger directivity is possible.

The use of smart antennas in sensor nodes is not only feasible, but also highly desirable. As sensor node dimensions shrink, RF communication will be forced to utilize higher frequencies. Fundamental theory states, however, that transmission using higher frequencies results in lower effective communication ranges. To compensate for distance loss, higher gains have to be achieved. Increased gains, which can be attained using smart antennas, are necessary to preserve connectivity in networks and efficiently use a sensor node's energy source. The advantages of using smart antennas in ad-hoc communications has been demonstrated using small-scale and large-scale fading models in [7] where improvements of 20dB in received signal to noise ratio (SNR) can be realized and the bit error rate can be reduced by more than 60%. Moreover, the use of smart antennas can significantly decrease the nodes' power consumption, and therefore increase their lifecycle. Consider for example the log-distance large-scale model for the channel path loss,

$$P_r = P_t G_r G_t \left(\frac{\lambda}{4\pi}\right)^2 \frac{1}{d^n}, \tag{1}$$

where where G_r and G_t are the antenna gains of the receiver and the transmitter respectively, P_r and P_t are the corresponding signal powers, λ is the wavelength of the electromagnetic signal, d is the distance between receiver and transmitter and n (typically between 1.7 and 6) is the power loss exponent of the channel.

Using smart antennas with higher gain than omni-directional antennas, the range of each node increases by

$$R_{dir} = R_{omni} \sqrt[n]{\left(\frac{G_{dir}}{G_{omni}}\right)^2}, \tag{2}$$

where R_{omni} and R_{dir} are the ranges achieved by omni-directional and smart antennas respectively, and G_{omni} and G_{dir} are the corresponding gains. Assuming switched beam antennas with perfect sectorization, then $G_{dir}/G_{omni} \approx 360/BW$. Additionally, if we want to cover the same range using smart antennas, the transmission power of each node will be reduced to,

$$P_{t_{dir}} = P_{t_{omni}} \left(\frac{G_{omni}}{G_{dir}}\right)^2, \tag{3}$$

regardless of the channel model's power ratio. For example, if both the transmitter and the receiver of a communication link use appropriately oriented antennas with 180 degrees beamwidth (BW), then the total transmit power needed for communication is equal to a quarter of the power that omni-directional antennas would need.

In the following sections we show the improvements in the performance of the network and how routing in sensor networks can be accomplished with minimal computations and power consumption when the proposed use of smart antennas is imposed.

3   Overview of Routing Protocols and Our Proposal

The problem of how to route data has been the subject of intense study in wireless sensor networks. Since these networks resemble mobile ad-hoc networks so closely, several MANET protocols have already been proposed for deployment. Popular routing solutions in such networks are DSDV [8], DSR [9], AODV [10] and TORA [11]. These protocols, however, are designed for networks with ID-based node addressing and are not considered efficient for sensor networks. A nice alternative constitutes attribute-based routing, where the final destination is identified by attributes such as location or sensor measurements. Sensor Protocols for Information via Negotiation (SPIN) [12] and Directed Diffusion [13] are two examples of attribute-based routing. In contrast with flat networks, hierarchical networks use clustering approaches to organize nodes into various levels of responsibilities. Then cluster-heads are selected to play the role of the coordinator of the other nodes in their clusters. LEACH (Low-Energy Adaptive Clustering Hierarchy) [14] and TTDD (Tow-Tier Data Dissemination) [15] are two such examples.

This abundance of routing protocols (of course the above list is by no means complete) suggests that efficient routing attempts to optimize a variety of different measures including efficiency, robustness, number of activated particles, etc. In our setting, however, where sensed data need to be sent to a receiving center, these approaches seem to overload the sensors' processors with unnecessary computations, as we only need to send a single message or packet back to the base station. Taking into account the small communication throughput and the limited memory and computational capabilities of sensor networks a simple flooding approach seems to be the best alternative.

Flooding is the most computationally efficient protocol due to its computational simplicity as every node broadcasts every new incoming packet. Therefore, data are bound to reach their destination, assuring correctness, and the protocol is immune to node failures, assuring robustness. Although this protocol can be integrated even in the most simplistic implementations of sensor nodes, it is extremely energy consuming as all nodes must receive and transmit the message at least once. Gossiping or wandering approaches [16] seem to alleviate this problem, at the cost, however, of increasing path lengths or failing to reach destination.

In this work we propose a new family of protocols that try maximizing efficiency and minimizing energy consumption by favoring certain paths of local data transmission towards the sink by using switched beam antennas at the nodes. Just like flooding, the protocol is very easy to be implemented as it only requires nodes to forward every new incoming packet. Unlike flooding however, it avoids depleting the network from its resources by restricting the nodes that receive and hence retransmit the message with the use of switched beam antennas.

Our Proposal

The mechanism that controls this propagation of information is the following; during the initialization phase of the network, the base station transmits a beacon frame with adequate power to be able to reach all the network's nodes. A conceptual representation of the proposed protocol is shown in Figure 1. Each

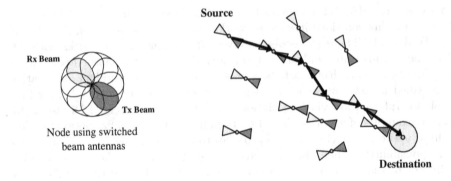

Fig. 1. Conceptual representation of the proposed protocol. All nodes have their transmit beams oriented towards the direction of the destination

node switches among its diverse beams and finds the one that delivers the best signal. After the initialization phase, the nodes will use this beam only for transmitting data, and they will use the beam lying on the opposite side of the plane only for receiving data. During normal operation, nodes retransmit every new incoming packet that has not received before. As it might be expected the protocol is highly dependent on the antenna beam-width (BW). By carefully selecting the appropriate beam-width one obtains a tradeoff between robustness (the fraction of times the message reaches the destination) and load incurred in the network (measured in terms of overall power consumption in the network). This is demonstrated in the following section.

4   Experimental Validation

In order to analyze the performance of the algorithm described above, we performed a set of large scale experiments whose goal was to test the protocol's effectiveness under the following measures:

1. *Success of delivery*: Messages should be delivered to the destination with high probability.
2. *Low activation of sensors*: A small number (compared to the total number) of nodes must be activated for each data transmission towards the sink.
3. *Power efficiency*: The overall network power consumption should be as small as possible. Although the number of activated nodes is a good measure it is not sufficient for our implementation since in the case of smart antennas the energy spent for receiving a packet is larger than that of transmitting one.
4. *Robustness under node failures*: The protocol should be able to deliver data to the destination, even when a large number of nodes is not responding. Here we analyze the protocol by incurring a failure or death probability on every sensor before the execution of the protocol.

5. *Scalability*: The performance of the routing algorithm should be independent of network size.
6. *Simplicity*: Any routing algorithm that is deployed in sensor networks must be able to run in an 8-bit microprocessor, with minimal data memory (i.e. 4Kbytes).

4.1   A Probabilistic Variant of the Protocol

The protocol described in the previous section is deterministic. This means that transmissions from the same source activate exactly the same set of nodes resulting in depletion of their energy. A way to reduce the utilization of the same sensors is to make the protocol probabilistic. A simple probabilistic variant is one where a sensor will retransmit a packet with probability depending on the power of the received signal. The closer this node is to a transmitting node, the greater the power of the received signal will be and hence the smaller the probability of retransmission. The intuition here is that nodes that are far from the sensor that is currently transmitting will retransmit with higher probability and hence information will reach the base station by using fewer nodes, and fewer hops. Hence the number of activated nodes will be reduced as well.

It should be noted that this new protocol uses *no distance* information (however, as we will show below the probability of retransmission is ultimately related to the distance of the receiver). At the receiver, the received signal strength varies from a maximum value $P_{r_{max}}$ to a minimum value $P_{r_{min}}$. These values depend on the sensitivity and the dynamic range of the receiver. According to the proposed protocol, the probability of retransmitting a received packet depends on the received signal strength. Therefore, when the received signal strength is close to $P_{r_{max}}$ then the probability of retransmitting the packet, $\Pr[Retransmit]$, will be close to zero. On the other hand, when the received signal strength is close to $P_{r_{min}}$, $\Pr[Retransmit]$ will be close to 1. In general, the probability of retransmitting the packet will vary linearly between these limits, depending on the received signal strength according to the equation

$$\Pr[Retransmit] = 1 - \frac{P_r - P_{r_{min}}}{P_{r_{max}} - P_{r_{min}}}, \tag{4}$$

where P_r is the received power measured at the receiver, and all values are measured in dB.

It turns out that the proposed probabilistic protocol is highly depended on the distance between the transmitter and the receiver of a single packet, and on the path loss power exponent. According to Equation (1) and assuming that $P_{r_{min}}$ is received at the maximum range R of a sensor node, then

$$P_{r_{min}} = P_t G_t G_r \left(\frac{\lambda}{4\pi}\right)^2 \frac{1}{R^n}. \tag{5}$$

From Equations (1), (4) and (5) it is evident that the probability of retransmitting a packet is equal to

$$\Pr[Retransmit] \propto \left(\frac{d}{R}\right)^n. \tag{6}$$

Thus the probability depends on the distance d between the transmitting and receiving station and on the path loss power exponent n. According to [17], n depends on the specific environment that the sensor network is deployed in, and it takes values between 1.9 and 5.4. For our simulations, we have chosen a characteristic value of $n = 3.3$, which is common for outdoor environments.

In the rest of the paper we will study the behavior of this probabilistic protocol under the metrics that were mentioned in the beginning of this section.

4.2   Invariance Under Network Size

It is obvious that the proposed protocol is extremely simple. Furthermore as it is demonstrated in Figure 2 it is scalable as its performance does not depend on network size but only on the average number of neighbors of each node. In this figure the activation and success ratios are shown as a function of the beamwidth for different network sizes ($N = 2500$, 5000 and 10000 nodes) provided the density (average number of neighbors μ) of each sensor remains the same, in this case $\mu = 10$, 25 and 50. As it can be clearly seen there is a very close match that essentially makes the behavior of the algorithm independent of network size.

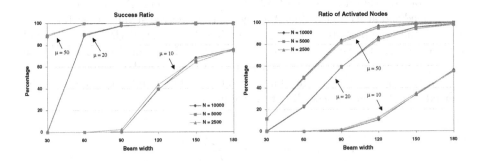

Fig. 2. Invariance under network size

Therefore, we will concentrate on proving the proposed protocol's power efficiency and effectiveness to reach destination. The effectiveness of the protocol is measured by the ratio of times the propagated event reached the sink while its efficiency is computed with respect to the number of activated nodes and the overall network power consumption. This last information is computed taking into account that a sensor's energy consumption depends on its communication range, its beamwidth, and the total number of packets it receives. Finally, as a case study, we compare the energy efficiency of our protocol against flooding and we show that the use of smart antennas can result in great savings. In order to evaluate our results, we used the sensor node model described in [18], which for completeness we present in Table 1.

Table 1. Node energy consumption model

Transmitter range:	1 to 10 m
Energy used	for reception: 30 nJ/bit
	for transmission: 20 nJ/bit + 1 pJ/bit/m^3
Packet sizes:	256 bits

Since we will also compare the proposed algorithm to the flooding mechanism that uses omni-directional antennas, the large-scale power model of Equations (1-3) can be used. According to Equation (3), the total power savings with respect to using omni-directional antennas at the nodes does not depend on the center frequency of the carrier signal or the path loss exponent of the channel. Therefore, the results of our simulations can be applied in all sensor network scenarios and for all wireless sensor node implementations that utilize a carrier frequency inside the RF or optical frequency spectrum.

We used the following setup for our experiments: $N = 10000$ sensors were spread uniformly at random in a square field of $10000m^2$, where all sensors have the same communication range R and switched beam antenna beam-width BW. We assume perfect sectorization of the beams and the same beam-width for transmit and receive beams. For each simulation run we choose the sensor with the smallest $x - y$ coordinates to be the source and the sensor with the largest $x - y$ coordinates to be the sink. In order to obtain valid statistical results all experiments were repeated a 1000 times. A typical run for $R = 3.5m$ and $BW = 30$ is shown in Figure 3.

4.3   Effectiveness

Figure 4 shows the success delivery ratio and activation percentage, when different ranges and different beam-widths are used. To have a reference measure, we include the correspondence between transmission range and average number of neighbors per node in the table that follows:

Range (in meters)	Average Density μ
1.8	10
2.2	15
2.9	25
4.0	50
5.8	100

In Figure 4(a) we see that the success ratio obeys a *threshold behavior*. So, for example, when $R = 5.8$ and $BW = 30$ the success probability is about 85% but

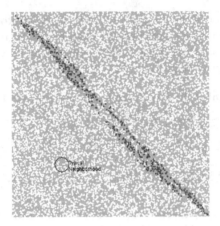

Fig. 3. Routing example for $N = 10000$, $R = 3.5m$ and $BW = 30$. Darker dots (blue) correspond to activated nodes (less than 5% in this case).

when the beam-width BW is 15, the success probability drops to zero. And the same behavior may be observed for all values of R. It is also obvious from this figure that the larger the communication range (or average number of neighbors) is the better the success probability becomes for any given beam width. Hence if we want to achieve a success ratio of 90%, we can either choose a BW of about 180 degrees when $R = 2.9$, a BW of 70 when $R = 4.0$, or a BW of 35 when $R = 5.8$. So, one may ask: are all these settings equivalent? The answer of course depends on the number of activated sensors which is shown on Figure 4(b).

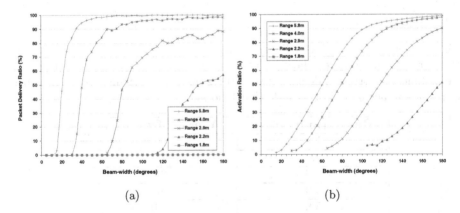

(a)   (b)

Fig. 4. (a) Success delivery ratio of proposed protocol. (b) Percentage of activated nodes

It can be seen in Figure 4(b) that given a specific range, the smaller the beam width is the lower the activation of sensors becomes. For example, to achieve a success ratio of 90% we see that we need about 90% of the sensors to be activated when R=2.9, while at R=5.8 less than 20% is activated to achieve the same result. However, one must be careful not to conclude from this figure that only antennas with small beam width and large transmission range should be used. For this conclusion to be valid, one must also take into account the power needed in order to transmit to a higher distance using a focused beam. This is something we do in a subsequent section.

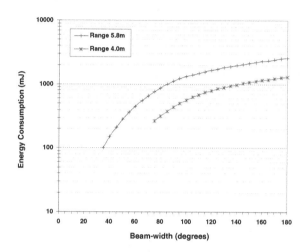

Fig. 5. Total network power consumption (mJ) in order to achieve PDR>90%

4.4   Robustness Under Failures

We also investigated the fault-tolerance nature of our protocol when sensors die with various probabilities. The invariance of our algorithm under changes in the network size suggests the following approach: when we know that sensors may die with certain probability we can either plant more nodes or increase the communication range slightly to counteract the effect of dead nodes. In any case, using the results of Figure 4, we can optimize the algorithm's performance and obtain the required robustness.

4.5   Power Efficiency

To compute the power consumption of the network, we apply Equations (1-3) to the node model of Table 1 and show the results on Figure 5. In general, we observe that for narrow beam-widths and large transmission ranges, the total

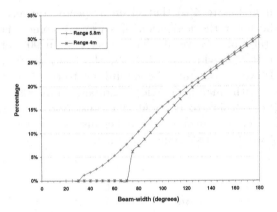

Fig. 6. Total network power consumption (%) with respect to flooding for PDR>90%

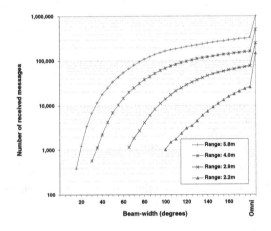

Fig. 7. Number of total received messages for the proposed algorithm and for flooding (Omni)

power that is consumed by the network in order to deliver a single packet is very small compared to using small transmission ranges and wide beam-widths. However, when we are allowed to work with antennas of fixed beam width then it is better to have a smaller transmission range. So for example when we restrict the beam-width to 80 degrees then in order to achieve packet delivery ratio (PDR)> 90% it makes sense to use a system that transmits at a range of 4.0m and not further.

Finally, Figure 6 shows that the power consumed by the network when the proposed algorithm is used is much smaller than that of flooding. The main reason for the evident power efficiency of the proposed algorithm is caused by the small number of nodes that are activated in the network compared to the total number of nodes that are transmitting when flooding is used. Moreover, due to the use of highly directional antennas, the total number of received messages is reduced, and therefore the total number of activated receivers is highly reduced as shown in Figure 7. Since, according to Table 1, the energy cost of transmitting a packet is comparable to that of receiving, the total energy consumed by the network using the proposed algorithm can be up to two orders of magnitude less than that of omni-directional flooding.

5   Conclusions and Future Research

In this work we have presented a routing algorithm for sensor networks that utilizes smart antenna systems, where sensed data is sent to a receiving center using only local information and total absence of coordination between sensors. Due to the novelty of our proposal we pointed out the feasibility and necessity of using smart antennas in sensor networks, as well as the advantages that are presented to communication links due to their use. Our protocol is suited for those cases where unexpected changes to the environment (i.e. a fire, a person entering a restricted area, etc.) must be propagated quickly back to the base station without the use of complicated protocols that may deplete the network from its resources. Our protocol is very easy to implement as nodes do not have to decide whether or not to forward the message. The protocol ensures packet delivery and low energy consumption solely with the use of smart antenna systems on sensor nodes.

We plan to continue this line of research by also considering "random" paths (not necessarily optimal) so that data is propagated to the destination. "Randomness" could be applied in the choice of the node beam direction, the transmission power, or the node antenna's beam-width (i.e with the use of variable gain switched beam antennas, like in [7]). Finally, mobility of sensors should be considered, so that networks using robotic sensors could be accounted for in the future. Of course, for this to be of any value, the protocol must again use only local information and no coordination between sensors.

References

1. I. F. Akyildiz, W. Su, Y. Sankarasubramaniam, and E. Cayirci, "Wireless sensor networks: a survey," *Computer Networks*, vol. 38, pp. 393–422, March 2002.
2. C.-Y. Chong and S. Kumar, "Sensor networks: evolution, opportunities and challenges," *Proceedings of the IEEE*, vol. 91, pp. 1247–1256, August 2003.
3. S. Tilak, N. Abu-Ghazaleh, and W. Heinzelman "A taxonomy of wireless microsensor network models". In *ACM Mobile Computing and Communications Review* (MC2R), Volume 6, 2, 2002.
4. B. Warneke, M. Last, B. Liebowitz, K.S.J. Pister, "Smart Dust: Communication with a Cubic-Millimeter Computer," *Computer*, vol. 34, pp. 44-51, January 2001.
5. B.A. Warneke, M.D. Scott, B.S. Leibowitz, Lixia Zhou, C.L. Bellew, J.A. Chediak, J.M. Kahn, B.E. Boser, Kristofer S.J. Pister, "An Autonomous $16mm^3$ Solar Powered Node for Distributed Wireless Sensor Networks", in *Proceeding of IEEE Sensors*,pp. 1510–1515, 2002.

6. Josep Diaz, Jordi Petit, Maria Serna, "A Random Graph Model for Optical Networks of Sensors", *IEEE Transactions on Mobile Computing*, vol. 3, pp. 186–196, 2003.
7. A. Kalis, Th. Antonakopoulos, "Direction finding in mobile Computing networks", *IEEE Transactions on Instrumentation and Measurement*, vol. 51, pp. 940–948, October 2002.
8. C. Perkins and P. Bhagwat, "Highly Dynamic Destination-Sequenced Distance-Vector Routing (DSDV) for Mobile Computers," in *Proceedings of the ACM SIGCOMM* 1994.
9. D. B. Johnson and D. A. Maltz, "Dynamic source routing in ad hoc wireless networks", *Mobile Computing*, Imielinski and Korth, Eds. Kluwer Academic Publishers, vol. 353, 1996.
10. C. E. Perkins and E. M. Royer, "Ad hoc On-Demand Distance Vector Routing", *Proceedings of the 2nd IEEE Workshop on Mobile Computing Systems and Applications*, New Orleans, LA, pp. 90-100, February 1999
11. V. D. Park and M. S. Corson, "A highly adaptive distributed routing algorithm for mobile wireless networks," *in Proceedings of IEEE INFOCOM*, vol. 3, pp. 1405–1413, April 1997.
12. J. Kulik, W. R. Heinzelman, and H. Balakrishnan "Negotiation-based Protocols for Disseminating Information in Wireless Sensor Networks", *ACM Wireless Networks*, vol. 8, 2002.
13. C. Intanagonwiwat, R. Govindan, D. Estrin, J. Heidemann, and F. Silva, "Directed diffusion for wireless sensor networking," *ACM/IEEE Transactions on Networking*, vol. 11, February 2002.
14. W. Heinzelman, A. Chandrakasan and H. Balakrishnan. "Energy-Efficient Communication Protocol for Wireless Microsensor Networks". HICSS 2000.
15. F. Ye, H. Luo, J. Cheng, S. Lu, and L. Zhang. "A two-tier data dissemination model for large-scale wireless sensor networks. In *Proc. of the 8th Inter. Conf. on Mobile Computing and Networking*, 2002.
16. Z. Haas, J. Y. Halpern, Li Li, "Gossip-Based Ad Hoc Routing", in *Proceedings of IEEE INFOCOM*, vol. 3, pp. 1707–1716, June 2002.
17. Sohrabi, K., Manriquez, B., Pottie, G.J., "Near ground wideband channel measurement", in *800-1000 MHz Vehicular Technology Conference*, 1999
18. R.C. Shah, J.M. Rabaey, "Energy aware routing for low energy ad hoc sensor networks", in *IEEE WCNC2002*, vol. 1, pp. 350–355, 2002

Neighborhood-Based Topology Recognition in Sensor Networks

S.P. Fekete[1], A. Kröller[1][*], D. Pfisterer[2][*], S. Fischer[2], and C. Buschmann[2]

[1] Department of Mathematical Optimization,
Braunschweig University of Technology,
D-38106 Braunschweig, Germany,
{s.fekete,a.kroeller}@tu-bs.de.
[2] Institute of Operating Systems and Computer Networks,
Braunschweig University of Technology,
D-38106 Braunschweig, Germany,
{pfisterer,fischer,buschmann}@ibr.cs.tu-bs.de.

Abstract. We consider a crucial aspect of self-organization of a sensor network consisting of a large set of simple sensor nodes with no location hardware and only very limited communication range. After having been distributed randomly in a given two-dimensional region, the nodes are required to develop a sense for the environment, based on a limited amount of local communication. We describe algorithmic approaches for determining the structure of boundary nodes of the region, and the topology of the region. We also develop methods for determining the outside boundary, the distance to the closest boundary for each point, the Voronoi diagram of the different boundaries, and the geometric thickness of the network. Our methods rely on a number of natural assumptions that are present in densely distributed sets of nodes, and make use of a combination of stochastics, topology, and geometry. Evaluation requires only a limited number of simple local computations.

ACM classification: C.2.1 Network architecture and design; F.2.2 Non-numerical algorithms and problems; G.3 Probability and statistics

MSC classification: 68Q85, 68W15, 62E17

Keywords: Sensor networks, smart dust, location awareness, topology recognition, neighborhood-based computation, boundary recognition, Voronoi regions, geometric properties of sensor networks, random distribution.

1 Introduction

In recent time, the study of wireless sensor networks (WSN) has become a rapidly developing research area that offers fascinating perspectives for combining technical progress with new applications of distributed computing. Typical scenarios

---

[*] Supported by DFG Focus Program 1126, "Algorithmic Aspects of Large and Complex Networks", Grants Fe 407/9-1 and Fi 605/8-1.

S. Nikoletseas and J. Rolim (Eds.): ALGOSENSORS 2004, LNCS 3121, pp. 123–136, 2004.
© Springer-Verlag Berlin Heidelberg 2004

involve a large swarm of small and inexpensive processor nodes, each with limited computing and communication resources, that are distributed in some geometric region; communication is performed by wireless radio with limited range. As energy consumption is a limiting factor for the lifetime of a node, communication has to be minimized. Upon start-up, the swarm forms a decentralized and self-organizing network that surveys the region.

From an algorithmic point of view, the characteristics of a sensor network require working under a paradigm that is different from classical models of computation: Absence of a central control unit, limited capabilities of nodes, and limited communication between nodes require developing new algorithmic ideas that combine methods of distributed computing and network protocols with traditional centralized network algorithms. In other words: How can we use a limited amount of strictly local information in order to achieve distributed knowledge of global network properties?

This task is much simpler if the exact location of each node is known. Computing node coordinates has received a considerable amount of attention. Unfortunately, computing exact coordinates requires the use of special location hardware like GPS, or alternatively, scanning devices, imposing physical demands on size and structure of sensor nodes. A promising alternative may be continuous range modulation for measuring distances between nodes, but possible results have their limits: The accumulated inaccuracies from local measurements tend to produce significant errors when used on a global scale. This is well-known from the somewhat similar issue of *odometry* from the more progressed field of robot navigation, where much more powerful measurement and computing devices are used to maintaining a robot's location, requiring additional navigation tools. See [SB03] for some examples and references. Finally, computation and use of exact coordinates of nodes tends to be cumbersome, if high accuracy is desired.

It is one of the main objectives of this paper to demonstrate that there may be a way to sidestep many of the above difficulties: Computing coordinates is not an end in itself. Instead, some structural location aspects do *not* depend on coordinates. An additional motivation for our work is the fact that location awareness for sensor networks in the presence of obstacles (i.e., in the presence of holes in the surveyed region) has received only little attention.

One key aspect of location awareness is *boundary recognition*, making sensors close to the boundary of the surveyed region aware of their position and letting them form connected *boundary strips* along each verge. This is of major importance for keeping track of events entering or leaving the region, as well as for communication purposes to the outside. Neglecting the existence of holes in the region may also cause problems in communication, as routing along shortest paths tends to put an increased load on nodes along boundaries, exhausting their energy supply prematurely; thus, a moderately-sized hole (caused by obstacles, by an event, or by a cluster of failed nodes) may tend to grow larger and larger.

Beyond discovering closeness to the boundary, it is desirable to determine the number and structure of boundaries, but also more advanced properties like membership to the *outer boundary* (which separates the region from the unbounded portion of the outside world) as opposed to the *inner boundaries* (which separate the swarm from mere holes in the region). Other important goals are the recognition of nodes that are well-protected by being far from the boundary, the recognition of nodes that are on the watershed between different boundaries (i.e., the Voronoi subdivision of the region), and the computation of the overall geometric *thickness* of the region, i.e., the size of the largest circle that can be fully inscribed into the region.

We show that based on a small number of natural assumptions, a considerable amount of location awareness can indeed be achieved in a large swarm of sensor nodes, in a relatively simple and self-organizing manner after deployment, without any use of location hardware. Our approach combines aspects of random distributions with natural geometric and topological properties.

Related Work. There are many papers dealing with node coordinates; for an overview, consider the cross-references for some of the following papers. A number of authors use anchors with known coordinates for computing node localization, in combination with hop count. See [DPG01,SRL02,SR02]. [CCHH01] uses only distances between nodes for building coordinates, based on triangulation. [PBDT03] presents a fully distributed algorithm that builds coordinate axes based on a by a near/far-metric and runs a spring embedder. Many related graph problems are NP-hard, as shown by [BK98] for unit disk graph recognition, and by [AGY04] for the case of known distances to the neighbors.

On the other hand, holes in the environment have rarely been considered. This is closely related to the k-coverage problem: Decide whether all points in the network area are monitored by at least k nodes, where k is given and fixed. In this context, a point is monitored by a node, if it is within the node's sensing area, which in turn is usually assumed to be a disk of fixed size. By setting the sensing range to half the communication range, 1-coverage becomes a decision problem for the existence of holes. In [HT03], an algorithm for k-coverage that can be distributed is proposed. Unfortunately, it requires precise coordinates for all nodes. In [FGG04], holes are addressed with greedy geographic routing in mind: Nodes where data packets can get stuck are identified using a fully local rule, allowing identification of the adjacent hole by using a distributed algorithm. Again, node coordinates must be known for both detection rule and bypassing algorithm to work. [Bea03] considers detection of holes resulting from failing nodes. It proposes a distributed algorithm that uses a hierarchical clustering to find a set of clusters that touch the failing region and circumscribe it.

Our Results. We show that distributed location awareness can be achieved without the help of location hardware. In particular:

- We describe how to recognize the nodes that are near the boundary of the region. The underlying geometric idea is quite simple, but it requires some effort on both stochastics and communication to make it work.
- We extend our ideas to distinguish the *outside boundary* from the *interior boundaries.*
- We describe how to compute both boundary distance for all nodes and overall region thickness.

– We sketch how to organize communication along the boundaries.
– We describe how to compute the *Voronoi boundaries* that are halfway between different parts of the boundary.

The rest of this paper is organized as follows. In Section 2 we give some basic notation and state our underlying model assumptions. In Section 3 we describe how to obtain an auxiliary tree structure that is used for computing and distributing global network parameters. Section 4 gives a brief overview of probabilistic aspects that are used in the rest of the paper to allow topology recognition. Section 5 describes how to perform boundary recognition, while Section 6 gives a sketch of how to compute more advanced properties. Section 7 describes implementation issues and shows some of our experiments. Finally, Section 8 discusses the possibilities for further progress based on our work.

2   Preliminaries

Swarm and geometry. In the following, we assume that the swarm consists of a set V of nodes, and the cardinality of V is some large number n. Each node $v \in V$ has a globally unique ID (for simplicity, denoted by v) of storage size $O(\log n)$, and coordinates x_v that are unknown to any node. All node positions are contained in some connected region $A_{\text{net}} \subset \mathbb{R}^2$, described by its boundary elements. In computational geometry, it is common to consider regions that are non-degenerate polygons, bounded by k disjoint closed polygonal curves, consisting of a total of s line segments, meeting pairwise in a total of s corners. Each of the k boundary curves separates the interior of A_{net} from a connected component of $\mathbb{R}^2 \setminus A_{\text{net}}$. The unique boundary curve separating A_{net} from the infinite component of $\mathbb{R}^2 \setminus A_{\text{net}}$ is called the *outside boundary*; all other boundaries are *inside boundaries*, separating A_{net} from *holes*. Thus, the genus of A_{net} is $k - 1$. As we do not care about the exact shapes (and an explicit description of the boundary is neither required nor available to the nodes), we do not assume that A_{net} is polygonal, meaning that curved (e.g., circular) instead of linear boundary pieces are admissible; we still assume that it consists of s elementary curves, joined at s corners, with a total number of k boundaries.

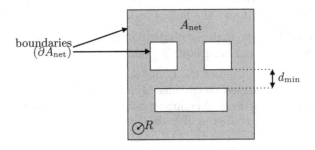

Fig. 1. Geometric parameters.

As narrow bottlenecks in a region can lead to various computational problems, a standard assumption in computational geometry is to consider regions with a lower bound on the *fatness* of the region; for a polygonal region, this is defined as the ratio between $d_{\min}$, the smallest distance between a corner of the region and a nonadjacent boundary segment, and $d_{\max}$, the diameter of the region. When dealing with sensor networks, the only relevant parameter for measuring distances is the communication radius, R. Thus, we we use a similar parameter, called *feature size*, which is the ratio between $d_{\min}$ and R. For the rest of this paper, we assume that feature size has a lower bound of 2. (This technical assumption is not completely necessary, but it simplifies some matters, which is necessary because of limited space.) In addition, we assume that angles between adjacent boundary elements are bounded away from 0 and from 2π, implying that there are no sharp, pointy corners in the region.

Node distribution. A natural scenario for the deployment of a sensor network is to sprinkle a large number of small nodes evenly over a region. Thus, we assume that the positioning of nodes in the region is the result of a random process with a uniform distribution on A_{net}. We also assume reasonable density; in a mathematical sense, this will be made clear further down. In a practical sense, we assume that each node can communicate with at least 100 other nodes, and the overall network is connected.

$\lambda(\cdot)$ denotes a volume function (i.e., the Lebesgue measure) on $\mathbb{R}^2$, therefore $0 < \lambda(A_{\text{net}}) < \infty$. For simplicity, $\lambda_{\circ} := \pi R^2$ denotes the area of the disk with radius R.

Using the notation $V(A) := \{v \in V : x_v \in A\}$, the expected number of nodes to fall into an area $A \subset A_{\text{net}}$ is therefore

$$\mathrm{E}[|V(A)|] = n\frac{\lambda(A)}{\lambda(A_{\text{net}})} \ . \tag{1}$$

Therefore, a node $v \in V$ that is not close to the network area's boundary, i.e., $B_R(x_v) \subset A_{\text{net}}$ has an estimated neighborhood size of

$$\mu := \mathrm{E}[|N(v)|] = (n-1)\frac{\lambda_{\circ}}{\lambda(A_{\text{net}})} \ . \tag{2}$$

Here, $B_r(x)$ denotes the ball around x with radius r.

Node communication. Nodes can broadcast messages that are received by all nodes within communication range. The cost of broadcasting one message of size m is assumed to be $O(m)$; e.g., any message containing a sender ID incurs a cost of $O(\log n)$.

We assume that two nodes $u \neq v \in V$ can communicate if, and only if, they are within distance R. This is modeled by a set of edges, i.e., $uv \in E : \iff \|x_u - x_v\| \leqslant R$, where $\|\cdot\|$ denotes the Euclidean norm. The set of adjacent nodes of $v \in V$ is denoted by $N(v)$, and does not include v itself. Such a graph is known under many names, e.g., geometric, (unit) disk, or distance graph. The maximum degree is denoted by $\Delta := \max_{v \in V} |N(v)|$.

3   Leader Election and Tree Construction

A first step for self-organizing the swarm of nodes is building an auxiliary structure that is used for gathering and distributing data. The algorithms that are presented in Section 5 only work if certain global network parameters are known to all nodes. By using a directed spanning tree, nodes know when the data aggregation phase terminates and subsequent algorithmic steps may follow. This is in contrast to other methods like flooding, where termination time is unknown. The issue of *leader election* has been studied in various contexts; see [BKKM96] for a good description. In principle, protocols for leader election may be used for our purposes, as they implicitly construct the desired tree; however, using node IDs (or pre-assigning leadership) does offer some simplification.

An alternative to leader election is offered by the seminal paper [GHS83] dealing with distributed and emergent search tree construction. It builds a minimum spanning tree in a graph with n nodes in a distributed fashion, using only local communication. Complexities are $O(n \log^2 n)$ time from the first message by a node to completed information at the constructed root, and $O(\log^2 n)$ transmissions per node, consisting of $O(\log n)$ messages, each sized $O(\log n)$.

In the following, we will use this auxiliary tree; in particular, we may assume that each node knows $N(v)$ and n, and it is able to use the tree for requesting and obtaining global data. Note that the tree is only used for bootstrapping the network; it may be replaced by a more robust structure at a later time.

4   Probabilistic Aspects

The idea for recognizing boundary nodes is relatively simple: Their communication range intersects a smaller than average portion of the region, and thus $N(v)$ is smaller than in other parts. However, a random distribution of nodes does not imply that the size of $N(v)$ is an immediate measure for the intersected area, as there may be natural fluctuations in density that could be misinterpreted as boundary nodes. In order to allow dealing with this difficulty, we introduce a number of probabilistic tools.

Recall that Chebychev's inequality shows that for a binomial distribution for n events with probability p, i.e., for a bin (n, p)-distributed random variable X, and $\alpha < 1$

$$\Pr[X \leqslant \alpha np] \leqslant \frac{1}{n} \cdot \text{const} \to 0 (n \to \infty) \tag{3}$$

holds. We exploit this fact to provide a simple local rule to let nodes decide whether they are close to the boundary ∂A_{net}. Let $\alpha < 1$ be fixed, and let

$$D = D(\alpha) := \{v \in V : |N(v)| \leqslant \alpha\mu\} . \tag{4}$$

Theorem 1. *Let v be a node whose communication range lies entirely in A_{net}. Then $v \notin D$ with high probability.*

Proof. This follows directly from (3), as

$$\Pr[|N(v)| \leqslant \alpha\mu] = \Pr[|V(B_R(x_v)) \setminus \{v\}| \leqslant \alpha\mu] \to 0 \quad (n \to \infty). \qquad (5)$$

□

Theorem 2. *Let* $x \in \partial A_{\mathrm{net}}$ *be on the network area's boundary. Let* $\varepsilon > 0$. *Assume* $\alpha > \frac{1}{\lambda_\circ}\lambda(B_{R+\varepsilon}(x) \cap A_{\mathrm{net}})$. *Then, with high probability, there is a node* $v \in D$ *with* $\|x - x_v\| \leqslant \varepsilon$.

Proof. Let $A_\varepsilon(x) := B_\varepsilon(x) \cap A_{\mathrm{net}}$ be the area where v is supposed to be. Then $\lambda(A_\varepsilon(x)) > 0$ by our assumption on feature size. The probability that there is no node in $A_\varepsilon(x)$ equals the probability for a bin $\left(n, \frac{\lambda(A_\varepsilon(x))}{A_{\mathrm{net}}}\right)$-distributed variable to become zero, i.e.,

$$\Pr[|V(A_\varepsilon(x))| = 0] = \left(1 - \frac{\lambda(A_\varepsilon(x))}{A_{\mathrm{net}}}\right)^n \to 0 \quad (n \to \infty). \qquad (6)$$

On the other hand, the probability that a node u in $A_\varepsilon(x)$ has more than $\alpha\mu$ neighbors is

$$\begin{aligned}
\Pr[|N(u)| > \alpha\mu] &= \Pr[|V(A_\varepsilon(x))| > \alpha\mu + 1 \mid u \text{ exists}] \\
&\leqslant \Pr[|V(B_{R+\varepsilon}(x))| > \alpha\mu + 1] \\
&\to 0 \quad (n \to \infty), \text{ because } \alpha\lambda_\circ > \lambda(A_{R+\varepsilon}(x)) .
\end{aligned}$$

Together, we get

$$\begin{aligned}
&\Pr[\exists v \in V, x_v \in A_\varepsilon(x) : |N(v)| \leqslant \alpha\mu] \\
&= 1 - \Pr[V(A_\varepsilon(x)) = \varnothing] - \Pr[\forall v \in V(A_\varepsilon(x)) : |N(v)| > \alpha\mu \mid V(A_\varepsilon(x)) \neq \varnothing] \\
&\geqslant 1 - \Pr[V(A_\varepsilon(x)) = \varnothing] - \Pr[|N(v)| > \alpha\mu \mid v \in V(A_\varepsilon(x))] \\
&\to 1 \quad (n \to \infty) ,
\end{aligned}$$

which proves the claim.

□

The assumed lower bound on α can be derived from natural geometric properties. For example, if all angles are between $\frac{\pi}{2}$ and $\frac{3\pi}{2}$, then for $\alpha > 0.75$ the condition holds for a reasonably small ε. We conclude that D reflects the boundary very closely. It can be determined by a simple local rule, namely checking whether the number of neighbors falls below $\alpha\mu$. However, this requires that all nodes know the value of $\alpha\mu$. The next Section 5 focuses on this key issue by providing distributed methods for estimating μ and α.

5   Boundary Computation

As described in the previous section, the key for deciding boundary membership is to obtain good estimates for the average density μ of fully contained nodes, and determining a good threshold α. In the following Subsection 5.1 we derive a method for determining a good value for μ. Subsection 5.2 gives an overview of the resulting distributed algorithm, if α has been fixed. The final Subsection 5.3 discusses how to find a good value for α.

5.1   Determining Unconstrained Average Node Degree μ

Computing the overall average neighborhood size can be performed easily by using the tree structure described above. However, for computing μ, we need the average over *unconstrained* neighborhoods; the existence of various pieces of boundary may lower the average, thus resulting in wrong estimates.

On the other hand, it is not hard to determine the *maximum* neighborhood size Δ. As was shown by [AR97], the ratio of maximum to average degree in the unit square intersection graph of a set of n random points with uniform distribution inside of a large square tends to 1 as n tends to infinity. We believe that a similar result can be derived for unit disk intersection graphs. Unfortunately, convergence of the ratio is quite slow, and using Δ as an estimate for μ is not a good idea. For our illustrative example with 45,000 nodes (see Figure 5), we get $\Delta/\mu \approx 1.37$.

However, even for very moderate sizes of n, Δ is within a small constant of μ, allowing us to compute the *node degree histogram* shown in Figure 2, again by using the auxiliary tree structure. Clearly, the histogram arises by overlaying three different distributions:

1. The neighborhood sizes of all non-boundary nodes.
2. The neighborhood sizes of near-boundary nodes, at varying distance from the boundary.
3. The neighborhood sizes of boundary nodes.

We expect a pronounced binomial distribution around μ for (1.), a uniform distribution for values safely between $\mu/2$ and μ for 2., overlayed with a small binomial distribution for values under $\mu/2$ for 3., possibly skewed in the presence of many nodes near corners of the region. (The latter is not to be expected under our geometric assumption of bounded feature size and minimum angle, but could be used as an indication of a large number of pointy corners otherwise.)

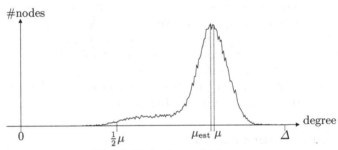

Fig. 2. Node degree histogram

Obviously, a variety of other conclusions could be drawn from the node degree histogram. Here we only use the most common neighborhood size μ_{est} as an estimate for μ. In our example, $\mu \approx 179.65$ and $\mu_{est} = 177$. The according histogram is shown in Figure 2. It resembles the expected shape very closely.

5.2 Algorithms

When the auxiliary tree is constructed, its root first queries the tree for Δ, and afterwards for the neighborhood size histogram. Using Δ, it can quantize the histogram to a fixed number of entries while expecting a high resolution. This step involves per-node transmissions of $O(1)$ for queries and $O(\log n)$ for the responses. On reception of the histogram, the root determines μ_{est}. Assuming that α is known, it then starts a network flood to pass the value $\alpha\mu_{est}$ to all nodes. Message complexity for the flood is $O(\log n)$.

A node receiving this threshold decides whether it belongs to D. In this case, it informs its neighbors of this decision after passing on the flood. These nodes form connected boundaries by constructing a tree as described in Section 3, with the additional condition that two nodes in D are considered being connected if their hop distance is at most 2.

The root of the resulting tree assigns the boundary a unique ID, e.g., its node ID. This ID is then broadcast over the tree. All nodes receiving their ID start another network flood. This flood is used such that each node determines the hop count to its closest boundary. If it receives messages informing it of two different boundaries at roughly the same distance, it declares itself to be a Voronoi node.

In addition, the boundary root attempts to establish a one-dimensional coordinate system in the boundary by sending a message token. The recipient of this token chooses a successor to forward the token to, which has to acknowledge this choice. Of the possible successors, i.e., not explicitly excluded nodes, the one having the smallest common neighborhood with the current token holder is chosen. The nodes receiving the token passing message without being the successor declare themselves as excluded for futher elections. After traveling a few hops, the boundary root gets prioritized in searching for the token's next hop, thereby closing the token path and forming a closed loop through the boundary. This path can then be used as axis for the one-dimensional coordinates.

5.3 Determining a Good Threshold α

Our algorithms depend on a good choice of the area-dependent parameter α, which should be as low as possible without violating the lower bound from Theorem 2. If a bound on corner angles is known in advance, say, $3\pi/2$ in a rectilinear setting, this is easy: For example, choose α slightly larger than $3/4$. As this may not always be the case, it is desirable to develop methods for the swarm itself to determine a useful α.

For a too small α, no node will be considered part of the boundary. For increasing α, the number of connected boundary pieces grows rapidly, until α is large enough to allow different pieces of the same boundary to grow together, eventually forming the correct set of boundary strips. When further increasing α, additional boundaries appear in low-density areas, increasing the number of identified boundaries. These boundaries also begin to merge, until eventually a single boundary consisting of the whole network is left.

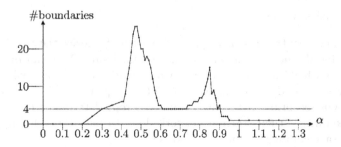

Fig. 3. The number of boundary components as a function of α.

Figure 3 shows that this expected behavior does indeed occur in reality: Notice the clear plateau at 4 connected components, embedded between two pronounced peaks.

This shows that computing a good threshold can be achieved by sampling possible values of α and keeping track of the number of connected boundary components.

6   Higher-Order Parameters

Once the network has identified boundary structures, it is possible to make use of this structure for obtaining higher-order information. In this section, we sketch how some of them can be determined.

6.1   Detection of Outer Boundary

One possible way to guess the outer boundary is to hope that the shape of boundary curves is not too complicated, which implies that the outside boundary is longest, and thus has the largest number of points. An alternative heuristic is motivated by the following theorem. See Figure 4 for the idea.

Theorem 3. *Let P be a simple closed polygonal curve with feature size at least $2R$ and total Euclidean length $\ell(P)$, consisting of edges e_i, and let φ_i be the (outside) angle between edges e_i and e_{i+1}. Let $B_i(P)$ be the set of all points that are near P as an inner boundary, i.e., that are outside of P and within distance R of P, and let $B_o(P)$ be the set of all points that are near P as an outside boundary, i.e., that are inside of P and within distance R of P. Then the area of $B_o(P)$ is $R\ell(P) - \sum_{\varphi_i>0} R^2 \frac{\varphi_i}{2} + \sum_{\varphi_i<0} R^2 \tan(\frac{-\varphi_i}{2})\pi R^2$, while the area of $B_i(P)$ is $R\ell(P) + \sum_{\varphi_i<0} R^2 \frac{\varphi_i}{2} - \sum_{\varphi_i>0} R^2 \tan(\frac{-\varphi_i}{2})\pi R^2$.*

Proof. As shown in Figure 4, both $B_o(P)$ and $B_i(P)$ can be subdivided into a number of strips s_i parallel to edges e_i of P, and circular segments near vertices of P that are either positive (when the angle between adjacent strips is positive, meaning there is a gap between the strips) or negative (when the angle is negative, meaning that strips overlap.) More precisely, if the angle φ_i between edges

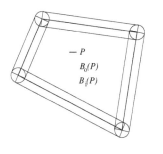
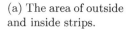
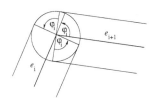

(a) The area of outside and inside strips.

(b) Overlap between adjacent strips.

Fig. 4. Geometry of boundary strips.

e_i and e_{i+1} is positive, we get an additional area of $R^2 \frac{\varphi_i}{2}$, while for a negative φ_i, the overlap is $R^2 \tan(\frac{\varphi_i}{2})$. In total, this yields the claimed area. Our assumption on feature size guarantees that no further overlap occurs.   □

Note that in any case, $\sum_i \varphi_i = 2\pi$. Making use of this property is possible in various ways: It is natural to assume that the area for the exterior boundary strip is less than suggested by its length, while it should be larger for all other boundaries. This remains true for other kinds of boundary curves using similar arguments.

A straightforward estimate for strip area is given by the number $|N(D_j)|$, while $|D_j|$ is a natural estimate for the length of the boundary, as the density of boundary nodes should be reasonably uniform along the boundary. Near boundary corners, the actual number of boundary nodes will be higher for convex corners (as the threshold of neighborhood size remains valid at larger distance from the boundary), and lower for nonconvex corners. Thus it makes sense to consider the ratio $\frac{|N(D_j)|}{|D_j|}$ for all boundary components, as the outside boundary can be expected to have lower than average $|N(D_j)|$ and higher than average $|D_j|$. The component with the lowest such ratio is the most likely candidate for being the outside boundary. See Table 1 for the values of our standard example.

Table 1. Number of boundary nodes and near-boundary nodes for each boundary component.

	D_1 (Outside boundary)	D_2 (Left eye)	D_3 (Right eye)	D_4 (Mouth)
$|N(D_j)|$	6093	1304	1319	2368
$|D_j|$	2169	289	266	616
$\frac{|N(D_j)|}{|D_j|}$	2.809	4.512	4.959	3.844

The main appeal of this approach is that the required data is already available, so evaluation is extremely simple; we do not even have to determine hop count along the boundary.

It should be noted that our heuristic may produce wrong results if there is an extremely complicated inside boundary. This can be fixed by keeping track of angles (or curvature) along the boundary; however, the resulting protocols become more complicated, and we leave this extension to future work.

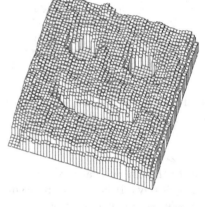

(a) Sensor network topology.   (b) Spatial distribution of $|N(v)|$.

Fig. 5. Example network consisting of 45,000 nodes.

6.2 Using Boundary Distance

Once all boundaries have been determined, it is easy to compute *boundary distances* for each node by determining a hop count from the boundary. Note that this can be done to yield non-integral distances by assigning fractional distances to the near-boundary nodes, depending on their neighborhood size.

This makes it easy to compute the geometric *thickness* of the region: Compute a node with maximum boundary distance. In our standard example, this node is located between the three inside boundaries.

7 Experimental Results

All our above algorithms have been implemented and tested on different point sets. See Figure 5(a) for an example with 45,000 nodes and four boundaries. The bounding box has a size of $30R \times 30R$. Total area of the region is $786.9R^2$. Figure 5(b) shows the spatial distribution of neighborhood size. Notice the slope near the boundaries. Figure 6(a) shows the identified boundary, near-boundary and Voronoi nodes, shown as black dots, gray crosses, black triangles, while

(a) Boundary nodes, near-boundary nodes, and Voronoi nodes.

(b) Structural loops along the boundaries.

Fig. 6. Experimental results for the example network.

other interior nodes are drawn as thin gray dots. The total number of identified boundary or near-boundary nodes is 11,358, leaving 33,642 nodes as interior nodes. Finally, Figure 6(b) shows the assigned structural loops along the various boundary strips.

8   Conclusions

We have shown that dealing with topology issues in a large and dense sensor network is possible, even in the absence of location hardware or the computation of coordinates. We hope to continue this first study in various ways. One possible extension arises from recognizing more detailed Voronoi structures by making use of the shape of the *boundary distance terrain*: This shape differs for nodes that have only one closest line segment in the boundary, as opposed to nodes that are close to two different such segment, constituting a ridge in the terrain. Note that our Voronoi nodes are close to pieces from two *different* boundaries.

An obvious limitation of our present approach is the requirement for high density of nodes. A promising avenue for overcoming this deficiency is to exploit higher-order information of the neighborhood structure, using more sophisticated geometric properties and algorithms. This should also allow the discovery and construction of more complex aspects of the network, e.g., for routing and energy management.

Acknowledgment. We thank Martin Lorek for his help.

References

[AGY04]   J. Aspnes, D. Goldenberg, and Y.R. Yang. On the computational complexity of sensor network localization. In *Proc. ALGOSENSORS*, 2004.

[AR97]   M.J.B. Appel and R.P. Russo. The maximum vertex degree of a graph on uniform points in $[0; 1]^d$. *Adv. Applied Probability*, 29:567–581, 1997.

[Bea03]   J. Beal. Near-optimal distributed failure circumscription. Technical Report AIM-2003-17, MIT Artificial Intelligence Laboratory, 2003.

[BK98]   H. Breu and D.G. Kirkpatrick. Unit disk graph recognition is NP-hard. *Comp. Geom.: Theory Appl.*, 9(1–2):3–24, 1998.

[BKKM96] J. Brunekreef, J.-P. Katoen, R. Koymans, and S. Mauw. Design and analysis of dynamic leader election protocols in broadcast networks. *Distributed Computing*, 9(4):157–171, 1996.

[CCHH01] S. Čapkun, M. Hamdi, and J. Hubaux. GPS-free positioning in mobile ad-hoc networks. In *Proc. IEEE HICSS-34—vol.9*, page 9008, 2001.

[DPG01]   L. Doherty, K.S.J. Pister, and L. El Ghaoui. Convex position estimation in wireless sensor networks. In *Proc. IEEE Infocom '01*, pages 1655–1663, 2001.

[FGG04]   Q. Fang, J. Gao, and L. J. Guibas. Locating and bypassing routing holes in sensor networks. In *Proceedings IEEE Infocom '04*, 2004.

[GHS83]   R. G. Gallager, P. A. Humblet, and P. M. Spira. A distributed algorithm for minimum-weight spanning trees. *ACM Transactions on Programming Languages and Systems*, 5(1):66–77, 1983.

[HT03]   C.-F. Huang and Y.-C. Tsent. The coverage problem in a wireless sensor network. In *Proc. ACM Int. WSNA*, pages 115–121, 2003.

[PBDT03] N.B. Priyantha, H. Balakrishnan, E. Demaine, and S. Teller. Anchor-free distributed localization in sensor networks. Technical Report MIT-LCSTR-892, MIT Laboratory for Computer Science, APR 2003.

[SB03]   C. Stachniss and W. Burgard. Mapping and exploration with mobile robots using coverage maps. In *Proc. IEEE/RSJ Int. Conf. IROS*, 2003.

[SR02]   N. Sundaram and P. Ramanathan. Connectivity-based location estimation scheme for wireless ad hoc networks. In *Proc. IEEE Globecom '02*, volume 1, pages 143–147, 2002.

[SRL02]   C. Savarese, J.M. Rabaey, and K. Langendoen. Robust positioning algorithms for distributed ad-hoc wireless sensor networks. In *Proc. 2002 USENIX Ann. Tech. Conf.*, pages 317–327, 2002.

A Novel Fault Tolerant and Energy-Aware Based Algorithm for Wireless Sensor Networks

Azzedine Boukerche[1*], Richard Werner Nelem Pazzi[2], and Regina B. Araujo[2]

[1] SITE – University of Ottawa
Ottawa, Canadá
boukerch@site.uottawa.ca
[2] DC - Universidade Federal de São Carlos, CP 676
13565-905 São Carlos, SP, Brazil
{richard,regina}@dc.ufscar.br

Abstract. Sensor networks are increasingly being deployed for finer grained monitoring of physical environments subjected to critical conditions such as fire, leaking of toxic gases and explosions. A great challenge to these networks is to provide a continuos delivery of events even in the presence of emergency conditions that can lead to node failures and path disruption to the sink that receives those events. A sensor network to cope with such situations has to be fast and reliable enough to respond to adversities. This paper presents a fault tolerant and low latency algorithm that meets sensors network requirements for critical conditions surveillance applications. The algorithm uses the publish/subscribe mechanism and the concept of driven delivery of events, a technique that selects the fastest path for the notification of events, reducing latency. Fault-tolerance is achieved through fast reconfiguration of the network, which switches from a shortest path mode to a multi-path reliable event notification mode.

1 Introduction

With the recent developments in wireless networks and multifunctional sensors with digital processing, power supply and communication capabilities, wireless sensor networks are being largely deployed in physical environments for fine-grained monitoring in different classes of applications. One of the most appealing applications is security surveillance and critical conditions monitoring. In a prison, for instance, it is important to keep a reliable monitoring of the physical environment, particularly when emergency situations emerge, such as a prisioners rebellion that can lead to incendiary fire conditions. In such situations, it is important that information can be sensed from the physical environment while the emergency state is in progress, since more precise information can be used by security and rescue teams for operation management and better strategic decisions. However, in order to keep the information flowing from the sensors during the emergency, a wireless sensor network solution

---

* Dr. A. Boukerche work was supported by NSERC and Canada Research Chair Program Grants

S. Nikoletseas and J. Rolim (Eds.): ALGOSENSORS 2004, LNCS 3121, pp. 137–146, 2004.
© Springer-Verlag Berlin Heidelberg 2004

has to cope with the failure of sensor nodes (sensors can be burnt, have their propagation jeopardized by interferences such as water or dense smoke present in the environment, can be malfunctioning etc). Thereby, wireless sensor network solutions for such environments have to be fault tolerant and reliable, and to provide low latency, besides fast reconfiguration and energy saving. In terms of energy savings, in a silent monitoring state, sensor nodes can be "programmed" to notify about events in a periodic (send temperature at every 10 minutes) or event-driven fashion (send temperature only when above 60 Deg. C). In these cases the interest may not change for quite some time. Some existing energy saving solutions take that into consideration and switch some nodes off, leading the nodes to an inactive state – these are waken up only when interest matches the events "sensed" [1] [2]. On the other hand, in query-based application scenarios, queries (new interests) can be propagated to sensors arbitrarily, according to the application and/or user´s will and so, some existing energy saving solutions may not be adequate because the transition from inactive state to data transfer state can be costly in terms of energy use when many arbitrary transitions are necessary [2]. Moreover, energy saving and fault tolerance can present conflicting interests when new paths, involving inactive nodes, have to be quickly set up because of failure in nodes of previous paths. This paper describes a wireless sensor network algorithm for information monitoring in critical conditions surveillance that can support simultaneously the three types of application scenarios described above by providing low latency event notification, fast broken paths reconfiguration, multiple alternative routes and energy saving. The publish/subscribe paradigm is used to promote the interaction between sensors and sink. Low latency is achieved by the use of the shortest path for the delivery of events. Fast subscriptions of new interests (for query-based scenarios) are provided by the concept of driven delivery of events, in which new subscriptions to a sensors region are speed up by using the inverse path used for event notifications. Fault tolerance is controled by the sink(s) - when the sink notices that abnormal events are being notified and that some nodes may be destroyed, it activates the multiway delivery, a network reconfiguration that changes sensor nodes to a mode that can set up multiple paths. The sensors network is configured through a hop tree, which is built at the configuration time. Subscriptions to the nodes are propagated to the sensors through the hop tree created. In order to better describe the algorithm, a grid model is used. However, the solution can be applied to mesh and dense randomly deployed sensor node networks as well.

2 Description of the Algorithm

The routing algorithm is realized in three steps. The first step comprises the construction of the hop tree. The sink starts the process of building the hop tree, which will be used as a configuration and subscription message propagation mechanism to the sensor network. The second step involves the propagation of subscriptions to the sensor network. The sink subscribes to the sensor network in order to receive particular information (events) from its nodes. Finally, the last step is responsible for delivering events from the sensors to the sink, by using the fastest and less costly route, in terms of energy savings. The routing steps are described in detail in the next sections. It is assumed that the nodes are disposed as a grid so that the transmission coverage of one sensor node is capable of reaching its eight neighbor nodes. However, the solution can be applied also for mesh networks model as well as dense randomly deployed sensor node networks, as shown along the text.

2.1 Building the Hop Tree

In the wireless sensor network considered here, one node does not have a global understanding of the network, i.e., a node only knows a small amount of information about its nearest neighbors (those that are within its coverage reach). In a first moment, each node knows only the hop level, of a hop tree, that it is in. The hop tree is started by a sink, which transmits to its neighbor(s) an attribute-value pair called hop. The algorithm for building the hop tree is based on flooding the network, starting from the sink, with a hop value, which is stored, incremented and transmitted to its neighbor nodes. These neighbor nodes store the received hop value, increment it and transmit it to its neighbor nodes and so on until the whole sensor network is configured with different levels of hops. Because the communication among the network nodes is through radio frequency, all the neighbors of a node receive the transmission. So, one node that has already transmitted, will receive its neighbor's transmission, generating a loop. In order to avoid these useless transmissions that causes energy waste, a set of rules was established as part of the algorithm for the hop diffusion. One of the local rules establishes that when a node receives a hop from its neighbor, it checks this value against its local hop value. If the local hop value is greater than the received one, the node updates its hop, increment this value and retransmit it to its neighbors. In case the locally stored hop is smaller or equal to the received hop, the node does not update its hop and does not transmit it. Thus, the network is configured, as shown in Figure 1, where each node knows only the hop level it is in. Figure 2 shows the same configuration for a mesh network.

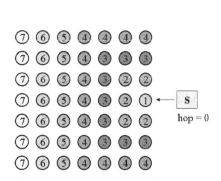

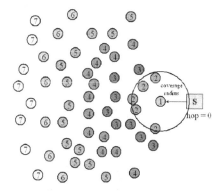

Fig. 1. Initial configuration of the network.   **Fig. 2.** Hop configuration in a mesh network.

The initial configuration algorithm is shown in Figure 3. The data structure used in the algorithm comprises three tables: configTable, routingTable and subscription-

Table. The configTable holds the configuration parameters associated with a sink. The routingTable is used by a node to forward messages to its neighboring nodes. Finally, the subscriptionTable is used to store subscriptions a node receives.

```
// configTable (hop, sinkID, subTimeStamp);
// routingTable (nID, sID, sinkID, coord);
// subscriptionTable (type, criteria, coord,
// sinkID, destID, senderID, timestamp, hop);
config.hop = 1;
config.sinkID = sinkID;
config.subTimeStamp = clock();
config.sendConfigMsg();
// When a node receives a configuration message,
// it checks its configTable to find a match.
Entry = configTable.get(config.sinkID);
if (entry)  // Entry exists?
{
    if (entry.hop > config.hop)
    {
        entry.hop = config.hop;
        config.hop = config.hop + 1;
        config.sendConfigMsg();
    }
}
else  // Entry does not exist!!
{
    entry.sinkID = config.sinkID;
    entry.hop = config.hop;
    configTable.add(entry);
    config.hop = config.hop + 1;
    config.sendConfigMsg();
}
```

Fig. 3. Initial Configuration algorithm and data structure.

2.2 The Subscription Message Propagation

In the publish/subscribe paradigm [3], for a sink to be notified about the events that are captured from the physical environment by the sensors, it needs to subscribe to one or more nodes for a given information, by setting one or more criteria (temperature > 60oC, presence of smoke, etc) that have to be matched before any event is sent. By sending events only when they match a criterion, it reduces network traffic, causing less waste of energy and extending the sensors network life. After the initial configuration of the network, the only information a node has is the hop level it is in. This information alone is not enough for the efficiency of the subscription propagation. In the absence of any information about which node of the network can satisfy the sink interest, one way to propagate the initial subscription is to flood the network with this interest. Each node of the network keeps a small subscription table and a routing table. Each record of the subscription table represents a different subscription. During the subscription message propagation, when a node receives this message, it compares the coord attribute to its own coordinates. If they are the same, it means that the subscription is meant to this node and so, it is stored in its subscriptions table. Otherwise, the node only re-transmits the subscription as part of the algorithm.

2.3 Sending the Notification Message

When information is captured from the physical environment by a sensor, it checks its subscription list to determine if there is any registered interest. If a criterion is met, the node verifies the senderID of the node that transmitted the subscription. After that, the node assembles an event notification message that contains the following attributes: type, value, coord, sinkID and send them to its neighbors. When each neighbor node receives the message, it compares the received destID with its own ID. If the result is true, the node stores coord and senderID in its routing table, gets the sID of the routing table and each node repeats the algorithm until the notification reaches the sink. Due to the initial configuration characteristic of the network, the maximum allowed number of neighbors that transmits to a node is three. Each node processes messages only from the nodes that are in an previous hop level. This characteristic makes it easy to select the neighbor that transmitted faster, besides avoiding messages loops. Supposing that a sink S sends a subscription to a network configured as shown in Figure 4, and considering that the top-left-most node is the sensor that produces an event that meets the subscription criterion of sink S, Figure 5 shows the path that is created down to the sink for the sending of the notification message.

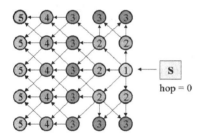

Fig. 4. Initial configuration and subscription propagation.

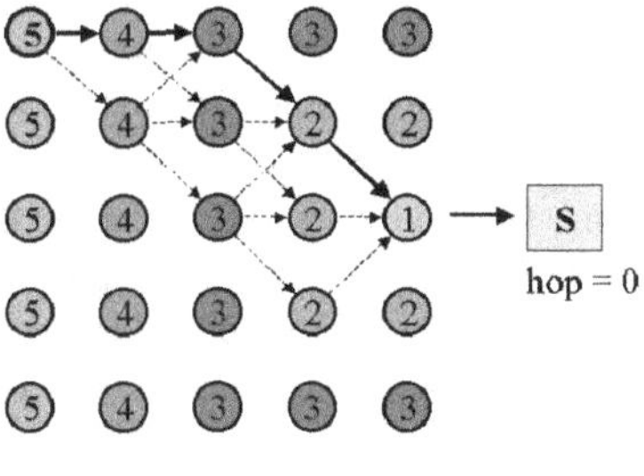

Fig. 5. Path created for the delivery of the notification.

Note that the arrows indicate the links that could form alternative paths, depending only on the choice each node makes for the fastest node that delivered the subscription. An important feature of the configuration of the hops values can be observed in the notification transmission phase. When a node receives a transmission from a neighbor node, it only retransmits the message if the node has higher hop number (one unity more). For instance, only the nodes with hop = 4 retransmit the information received from nodes with hop = 5, and so on. The pseudo code for the notification algorithm is shown in Figure 6. It can be seen from Figure 5 that the nodes that are not part of the path to the sink do not have arrows. This means that these nodes do not transmit any message at that point and so, do not waste as much energy as the others that are part of the path.

```
// When a node receives a notification message
if (notif.destID == node.ID)   // Is this the destination?
{
        // Gets the record for sinkID.
      route = routingTable.get(notif.sinkID);
        // Stores the sender ID in the routing table.
      route.nID = notif.senderID;
        // Stores the coord in the routing table.
      route.coord = notif.coord;
        // Sets the destination.
      notif.destID = route.sID;
        // Sends the message.
      notif.sendNotificationMsg();
}
```

Fig. 6. Notification Algorithm

2.4 Driven Delivery of Subscription Messages

The main goal of the notification messages propagation algorithm is to find the fastest path between the node that produced the event and the sink. The path that is used to deliver notifications can be used later by the sink to send new subscriptions to the same region which is delimited by the attribute coordinates. For that, during the delivery of the notification, each node in the path to the sink stores, in the routing table, the attribute coord, as well as the senderID of the neighbor node that transmitted the notification message. When a node receives a subscription message, it compares the attribute coord of the subscription with the stored coord. If the result is true, the node retransmits the message specifying the destID attribute so that only the neighbor node that matches this destID is allowed to transmit. Thus, only the nodes that were part of a route to previous notifications will transmit. In summary, in order to speed up new subscriptions to a sensors region, the subscription step can use the inverse path used for notification, as shown in Figure 7 (named as driven delivery).

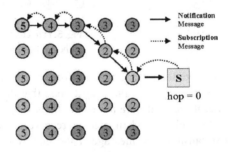

Fig. 7. Driven Delivery of Subscriptions.

This is useful when subscriptions of query-driven type have to be supported. Otherwise, if the node does not have a matching value for coord, the node transmits without specifying the senderID, so that all neighbors will transmit the subscription message. According to this algorithm, one node transmits only if its hop value matches the hop value received. The subscription message delivery algorithm is shown in Figure 8.

```
// When a node receives a subscription msg, it first
// checks to see if its hop value matches the
// subscription hop value.
// Gets the config record associated with sinkID.

config = configTable.get(sub.sinkID);
if (sub.hop == config.hop)
    // To get only the first subscription msg.
  if (config.subTimeStamp <> sub.timeStamp)
    // checks to see if this node is the publisher.
    if (node.coord == sub.coord AND node.type == sub.type)
    {       // Does this subscription exist?
        if (subscriptionTable.match(sub))
            subscriptionTable.refresh(sub);
        else  // it stores the subscription.
            subscriptionTable.add(sub);
    }
    else
    {       // checks if it is the destination.
        if (node.ID == sub.destID)
        {    // Tries to get the record
            route.routingTable.get(sub.coord)
            if (route) // Is there a route?
              sub.destID = route.nID;//sets dest.
            else // There is no route!
            {       // Stores the sender ID.
                route.sID = sub.senderID;
                // To send its ID to the other nodes.
                sub.senderID = node.ID;
                // Msg destined to all neighbors.
                sub.destID = null;
                // Only nodes hop+1 process the msg.
                sub.hop = sub.hop+1;
                sub.sendSubMsg(); // Sends the msg.
            }
        }
        else  // msg destined to all neighbors?
            if (sub.rID == null)
            { // To avoid get same subscript from other nodes.
                config.subTimeStamp = sub.timeStamp;
                // Used to forward notific msgs.
                route.sID = sub.senderID;
                route.sinkID = sub.sinkID;
                // Add the route to the routing table.
                routingTable.add(route);
                // Only nodes hop+1 process the msg.
                sub.hop = sub.hop+1;
                sub.sendSubMsg(); // Sends the msg.
            }
    }
```

Fig. 8. Subscription algorithm.

Because the driven delivery of subscriptions use the same path created for notification messages, only the nodes comprising this path spend energy for transmission. The other nodes either receive and do not transmit (as is the case of the neighbor nodes to the path nodes) or do not even receive messages. Figure 9 shows a map that represents the energy consumed by the network when using the referred path. The darker nodes denote a larger expenditure of energy.

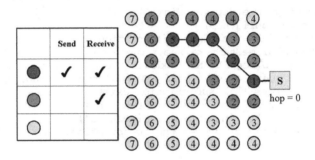

Fig. 9. Energy map of the network.

2.5 Repairing Broken Paths and Setting Up Multiple Paths for Fault Tolerance

The path created for sending the notification message is unique and the most efficient (promotes lower latency and saves energy). It can also be used for the driven delivery of new subscriptions (for query-driven scenarios, for instance, that may require random subscriptions). However, because the path is unique, any failure in one of its nodes will cause disruption, preventing the delivery of the event as well as the subscriptions. Possible causes of failure include: low energy, physical destruction of one or more nodes, communication blockage, etc. When a sink notices that the network needs reconfiguration because of path disruptions, it resends a configuration message, assigning new values to the nodes. A disrupted path is shown in Figure 10. When the sink resends the configuration message, the path for the notification messages can be rebuilt, as shown in Figure 11. Obviously, if all neighbors of a node fail this node will be isolated and its transmission will not reach any node. One solution would be to configure the radio module of the node to increase its coverage area, but this will spend more energy. Another solution is to provide faul tolerance through the establishment of multiple paths from the nodes to the sink. When a sink receives a critical event, such as fire detection, it can inform the sensor network that from now on the nodes can use all their neighbors to deliver events to the sink. This mechanism is necessary due to the greater probability of node's failures in a situation of fire or in other critical circunstances. The nodes may switch to an emergency mode where each node transmits the information specifying the receivers to be only the nodes with a one-unit less hop level.

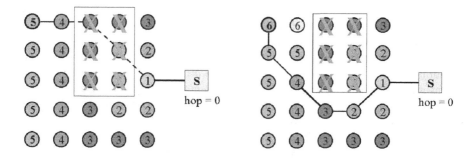

Fig. 10. Region with destroyed nodes.   **Fig. 11.** Reconfiguration of the network.

For instance, a node on hop level 5 will specify the receivers to be neighbor nodes on hop level 4. This way, more paths may be created to deliver the notification message to the sink, and thus a greater reliability can be achieved, as we can see in Figure 12. Also note that a node may receive the same message from distinct neighbors. To avoid this message repetition, a receiving node can verify its recent transmission cache to see if it has already received the message and decide not to transmit it. The decision about reconfiguration of the network is always from the sink. A sink may decide to reconfigure the network when enough abnormal events are received, in order to change it to emergency mode, where multiple paths are set up for notification delivery. Ambiguity situations are also dealt with at the sink, ie, the sink decides, from the established policies, if an abnormal event received from one sensor is an event that mirrors what is happening in the physical environment or if it is only the result of a malfunctioning sensor.

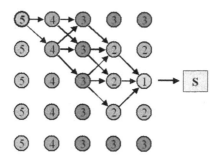

Fig. 12. Fault tolerance through the emergency mode delivery mechanism

3   Related Work

It is well known that a sensor in idle state consumes almost the same amount of energy than when it is active and energy savings means to turn off communications completely [2]. STEAM [5] provides a good solution for energy saving when the sensors have to be switched to the data path once in a while, ie., when application scenario is basically event-driven. However, when the switch has to be made often, such as when different types of subscriptions are requested (query-based), then the switch could outperform the savings of energy. In the algorithm described here, the three application scenarios are supported (periodic, event-driven and query-based). Other solutions are very efficient in energy saving, but do not support fault tolerance, such as EAD [4] e LEACH [6]. In other paradigms, like Directed Diffusion [1], every node stores every interest message, even if the node does not publish a matching event. In our algorithm, each intermediate node has a routing table to direct incoming messages, and it does not have to run a complete matching algorithm every time it receives a message.

4 Conclusions

This paper describes a wireless sensor network algorithm for information monitoring in critical conditions surveillance. Low latency is achieved by the use of the shortest path for the delivery of events. Fast subscriptions of new interests (for query-based scenarios) are provided by the concept of *driven delivery of events,* in which new subscriptions to a sensors region are speed up by using the inverse path used for event notifications. Fault tolerance is controlled by the sink(s) so that, when the sink notices that abnormal events are being notified and that some nodes may be destroyed, it activates a multi-path delivery, a reconfiguration of the network that changes the sensors nodes to a mode that can set up multiple paths.

References

1. C. Intanagonwiwat, R. Govindan and D. Estrin: Directed Diffusion: A Scalable and Robust Communication Paradigm for Sensor Networks. In Proc. 6th ACM/IEEE International Conference on Mobile Computing – MOBICOM'2000.
2. I. Chatzigiannakis, S. Nikoletseas and P. Spirakis: A Comparative Study of Protocols for Efficient Data Propagation in Smart Dust Networks. In Proc. 2nd ACM Workshop on Principles of Mobile Computing – POMC'2002.
3. P. T. Eugster, P. Felber, R. Guerraoui, A. Kermarrec: The many faces of publish/subscribe. ACM Comput. Survey. 35(2): 114-131 (2003)
4. A. Boukerche, X. Cheng, J. Linus, Energy-Aware Data-Centric Routing in Microsensor Networks. In MSWiM'03, September 19, 2003, San Diego, California, USA. (2003).
5. C. Schurgers, V. Tsiatsis, S. Ganeriwal and M. Srivastava: Topology Management for Sensor Networks: Exploiting Latency and Density. In Proc. MOBICOM 2002.
6. W. R. Heinzelman, A. Chandrakasan and H. Balakrishnan: Energy Efficient Communication Protocol for Wireless Microsensor Networks. In Proc. 33rd Hawaii International Conference on System Sciences – HICSS'2000.

Route Discovery with Constant Memory in Oriented Planar Geometric Networks

E. Chávez[1], S. Dobrev[2], E. Kranakis[3], J. Opatrny[4], L. Stacho[5], and J. Urrutia[6]

[1] Escuela de Ciencias Fisico-Matemáticas de la Universidad Michoacana de San Nicolás de Hidalgo, México.
[2] School of Information Technology and Engineering (SITE), University of Ottawa, 800 King Eduard, Ottawa, Ontario, Canada, K1N 6N5. Research supported in part by NSERC (Natural Science and Engineering Research Council of Canada) grant.
[3] School of Computer Science, Carleton University, 1125 Colonel By Drive, Ottawa, Ontario, Canada K1S 5B6. Research supported in part by NSERC (Natural Science and Engineering Research Council of Canada) and MITACS (Mathematics of Information Technology and Complex Systems) grants.
[4] Department of Computer Science, Concordia University, 1455 de Maisonneuve Blvd West, Montréal, Québec, Canada, H3G 1M8. Research supported in part by NSERC (Natural Science and Engineering Research Council of Canada) grant.
[5] Department of Mathematics, Simon Fraser University, 8888 University Drive, Burnaby, British Columbia, Canada, V5A 1S6. Research supported in part by NSERC (Natural Science and Engineering Research Council of Canada) grant.
[6] Instituto de Matemáticas, Universidad Nacional Autónoma de México, Área de la investigación cientifica, Circuito Exterior, Ciudad Universitaria, Coyoacán 04510, México, D.F. México

Abstract. We address the problem of discovering routes in strongly connected planar geometric networks with directed links. We consider two types of directed planar geometric networks: Eulerian (in which every vertex has the same number of ingoing and outgoing edges) and outerplanar (in which a single face contains all the vertices of the network). Motivated by the necessity for establishing communication in wireless networking based only on geographic proximity, in both instances we give algorithms that use only information that is geographically local to the vertices participating in the route discovery.

1 Introduction

The most extensively used model of wireless network has bidirectional links in the sense that packets may flow in any direction between any pair of adjacent vertices. This type of connectivity arises by considering a network in which all wireless hosts have the same transmission power, thus resulting in identical reachability radii. Such a network can be represented by a unit disk graph model.

A most important consideration is how to discover a route (among the many potential candidates) that must accommodate three rather contradictory goals: (1) avoiding flooding the network, (2) efficiency of the resulting path, (3) using only geographically local information. Although it is generally accepted that

S. Nikoletseas and J. Rolim (Eds.): ALGOSENSORS 2004, LNCS 3121, pp. 147–156, 2004.

flooding must be avoided in order to increase the network lifetime, limitations on the knowledge of the hosts make it necessary that in practice one may need to drop the second goal in favour of the third. Indeed, this is the case in geometric planar networks whereby algorithms are given for discovering routes (see Kranakis et al. [8]). Another important consideration is to construct a planar geometric network from a rather complicated wireless network. This question is addressed in Bose et al. [4] as well as in several subsequent papers. The basic idea is to preprocess the wireless network in order to abstract a planar geometric network over which the algorithm in Kranakis et al. [8] (compass routing, face routing, and other related routing algorithms) can be applied.

1.1   A Fundamental Issue

In principle, routing must be preceded by a radiolocation based vertex discovery process relying on an available Geographic Positioning System (or GPS) that will enable vertices to discover their neighbors. Although vertex discovery is not the purpose of study of this paper it is important to note that bidirectional communication will be rarely valid in practice. This may be due to several factors, including obstacles that may either obstruct direct view to a host and/or diminish the strength of a signal during propagation or even wireless hosts with different power capabilities.

A fundamental issue is whether geographically local routing is possible in arbitrary graphs. If the underlying network is either not planar or its links are not bidirectional the techniques outlined above may fail to route and/or traverse the network using only "geographically" local information. In general this is expected to be a difficult problem because one will never be able to avoid "loops" based only on geographically local information. Therefore there is a need for reexamination of the structure of the underlying backbone network that gives rise to our basic communication model in order to accomplish routing satisfying the previously set conditions.

1.2   Related Literature

There has been extensive literature related to discovering routes in wireless ad-hoc networks when the underlying graph is a undirected planar geometric network, e.g., see Bose et al. [4], Kranakis et al. [8], Kuhn et al. [9,10]. A problem related to routing is traversal which is addressed in several papers Avis et al. [1], Bose et al. [3], Chavez et al. [5], Czyczowicz et al. [6], Gold et al. [7], Peuquet et al. [11,12]. However, traversing all the vertices of a graph may be an "overkill" especially if all one requires is a single path from a source to a destination host.

1.3   Results of the Paper

In this paper we address the problem of discovering routes in strongly connected planar geometric networks with directed links using only local information. After clarifying the model in Section 2 we consider two types of directed planar geometric networks. In Section 3 we look at Eulerian planar geometric networks in which every vertex has the same number of ingoing and outgoing edges. In Section 4 we investigate outerplanar geometric networks whereby a single face contains all the vertices of the network.

2   Model

A *planar geometric network* is a planar graph G with vertex set V, edge set E and the face set F together with its straight line embedding into the plane $\mathbb{E}^2$. In this paper we always consider only finite graphs. Furthermore, we assume that no edge passes through any vertex except its end-vertices. A geometric network is *connected* if its graph is connected. An *orientation*   of a planar geometric network G is an assignment of a direction to every edge e of G. For an edge e with endpoints u and v, we write $e = (u, v)$ if its direction is from u to v. The geometric network together with its orientation is denoted by $\boldsymbol{G}$.

We assume that every vertex v is uniquely determined by the pair $[x, y]$ where x is its horizontal coordinate and y is its vertical coordinate.

Consider a connected planar geometric network $\boldsymbol{G} = (V, E)$. We say $\boldsymbol{G}$ is *Eulerian* if for every vertex $u \in V$, the size of $N^+(u) = \{x, (u, x) \in E\}$ equals the size of $N^-(u) = \{y, (y, u) \in E\}$; i.e. the number of edges outgoing from u equals the number of edges ingoing into u.

3   Route Discovery in Eulerian Planar Geometric Networks

First consider a planar geometric network G without any orientation. Given a vertex v on a face f in G, the boundary of f can be traversed in the counter-clockwise (clockwise if f is the outer face) direction using the well-known right hand rule [2] which states that it is possible to visit every wall in a maze by keeping your right hand on the wall while walking forward. Treating this face traversal technique as a subroutine, Kranakis et al. [8] give an elegant algorithm for routing in a planar geometric network from a vertex s to a vertex t.

If we impose an orientation   on G, then this algorithm will not work since some edges may be directed in an opposite direction while traversing a face. In this section, we describe a simple technique on how to overcome this difficulty. In particular, we propose a method for routing a message to the other end of an oppositely directed edge in Eulerian geometric networks.

Now suppose   is so that $\boldsymbol{G}$ is an Eulerian planar geometric network. For a given vertex u of $\boldsymbol{G}$, we order edges (u, x) where $x \in N^+(u)$ clockwise around u starting with the edge closest to the vertical line passing through u. Similarly we order edges (y, u) where $y \in N^-(u)$ clockwise around u; see Figure 1. Clearly, this orderings are unique and can be determined locally at each vertex.

Let $e = (y, u)$ be the i-th ingoing edge to u in $\boldsymbol{G}$. The function $\mathbf{succ}(e)$ will return a pointer to the edge (u, x) so that (u, x) is the i-th outgoing edge from

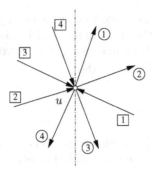

Fig. 1. Circled numbers represent the ordering on outgoing edges, squared numbers on ingoing ones.

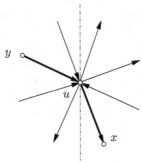

Fig. 2. In this example the ingoing edge (y, u) is third, so the chosen outgoing edge (u, x) is also third. Both these edges are depicted bold.

u. For an illustration of the function see Figure 2. Again, this function is easy to implement using only local information.

Obviously, the function **succ**() is injective, and thus, for every edge $e = (u, v)$ of G, we can define a closed walk by starting from $e = (u, v)$ and then repeatedly applying the function **succ**() until we arrive at the same edge $e = (u, v)$. Since G is Eulerian, the walk is well defined and finite. We call such a walk a *quasi-face* of G.

The following is the route discovery algorithm from [8] for planar geometric networks. We modify it so that it will work on Eulerian planar geometric network. For this, we only need to extend the face traversal routine as follows:

Whenever the face traversal routine wants to traverse an edge $e = (u, v)$ that is oppositely directed, we traverse the following edges in this order:

$$\mathbf{succ}(e), \mathbf{succ}(e)^2, \dots, \mathbf{succ}(e)^k,$$

so that $\mathbf{succ}(e)^{k+1} = (u, v)$. After traversing $\mathbf{succ}(e)^k$, the routine resumes to the original traversal of the face.

This modification obviously guarantees (in Eulerian geometric networks) that all edges of the face will be visited.

Algorithm 1   Eulerian Geometric Network Route Discovery.
Input: Connected Eulerian geometric network $G = (V, E)$
Starting vertex: s
Destination vertex: t

1: $v \leftarrow s$ {Current vertex = starting vertex.}
2: **repeat**
3:   Let f be a face of G with v on its boundary that intersects the line v-t at
     a point (not necessarily a vertex) closest to t.
4:   **for all** edges xy of f **do**
5:     **if** $xy \cap v$-$t = p$ and $\mathbf{dist}(p, t) < \mathbf{dist}(v, t)$ **then**
6:       $v \leftarrow p$
7:     **end if**
8:   **end for**
9:   Traverse f until reaching the edge xy containing the point p.
10: **until** $v = t$

Theorem 1. *Algorithm 1 will reach t from s in at most $O(n^2)$ steps.*

Proof. Follows from the proof of the correctness of the traversal algorithm for planar geometric networks from [8] and from the discussion above. The bound on the number of steps follows from the $O(n)$ bound on the number of steps of the algorithm from [8] and the fact that every edge on the route can be oppositely oriented and may need up to $O(n)$ steps to route through.

Examples which show that the bound cannot be improved in general are easy to construct, they typically include a large face that needs to be traversed whose boundary contains $\Theta(n)$ edges which are all oriented the opposite way.

4   Route Discovery in Strongly Connected Outerplanar Geometric Networks

A planar geometric network G is *outerplanar* if one of the elements in F contains all the vertices—the outerface. We will assume that this face is a convex polygon in $\mathbb{E}^2$. For a given triple of vertices x, y, and z, let $V_\curvearrowright(x, y, z)$ [$V_\curvearrowleft(x, y, z)$, resp.] denote the ordered set of vertices distinct from x and z that are encountered while moving from y counterclockwise [clockwise, resp.] around the outerface of G until either x or z is reached; see Figure 3.

Now consider an orientation   of the geometric network G. Let $N_\curvearrowright(x, y, z) = V_\curvearrowright(x, y, z) \cap N^+(x)$ and let $N_\curvearrowleft(x, y, z) = V_\curvearrowleft(x, y, z) \cap N^+(x)$. If $N_\curvearrowright(x, y, z) \neq \emptyset$, let $v_\curvearrowright(x, y, z)$ denote the first vertex in $N_\curvearrowright(x, y, z)$. Similarly we define $v_\curvearrowleft(x, y, z)$ as the first vertex in $N_\curvearrowleft(x, y, z)$, if it exists. A geometric network with fixed orientation is *strongly connected* if for every ordered pair of its vertices, there is a (directed) path joining them.

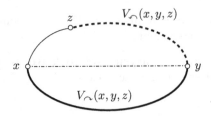

Fig. 3. The dashed part of the outer face represents the vertices in $V_\frown(x, y, z)$ and the bold solid part represents vertices in $V_\frown(x, y, z)$, respectively. Note that y belongs to both these sets and is in fact the first element of those sets.

Algorithm 2   Outerplanar Geometric Network Route Discovery.

Input: Strongly connected outerplanar geometric network $G = (V, E)$
Starting vertex: s
Destination vertex: t

1: $v \leftarrow s$ {Current vertex = starting vertex.}
2: $v_\frown, v_\frown \leftarrow s$ {counterclockwise and clockwise bound = starting vertex.}
3: **while** $v \neq t$ **do**
4:     **if** $(v, t) \in E$ **then**
5:        $v, v_\frown, v_\frown \leftarrow t$ {Move to t.}
6:     **else if** $N_\frown(v, t, v_\frown) \neq \emptyset$ and $N_\frown(v, t, v_\frown) = \emptyset$ **then** {No-choice vertex; greedily move to the only possible counterclockwise direction toward t.}
7:        $v, v_\frown \leftarrow v_\frown(v, t, v_\frown)$
8:     **else if** $N_\frown(v, t, v_\frown) = \emptyset$ and $N_\frown(v, t, v_\frown) \neq \emptyset$ **then** {No-choice vertex; greedily move to the only possible clockwise direction toward t.}
9:        $v, v_\frown \leftarrow v_\frown(v, t, v_\frown)$
10:    **else if** $N_\frown(v, t, v_\frown) \neq \emptyset$ and $N_\frown(v, t, v_\frown) \neq \emptyset$ **then** {Decision vertex; first take the "counterclockwise" branch but remember the vertex for the backtrack purpose.}
11:       $b \leftarrow v$; $v, v_\frown \leftarrow v_\frown(v, t, v_\frown)$
12:    **else if** $N_\frown(v, t, v_\frown) = \emptyset$ and $N_\frown(v, t, v_\frown) = \emptyset$ **then** {Dead-end vertex; backtrack to the last vertex where a decision has been made. No updates to $v_\frown$ and $v_\frown$ are necessary.}
13:       **if** $v \in V_\frown(t, b, t)$ **then**
14:          **while** $v \neq b$ **do**
15:             $v \leftarrow v_\frown(v, b, v)$
16:          **end while**
17:       **end if**
18:       **if** $v \in V_\frown(t, b, t)$ **then**
19:          **while** $v \neq b$ **do**
20:             $v \leftarrow v_\frown(v, b, v)$
21:          **end while**
22:       **end if**
23:       $v, v_\frown \leftarrow v_\frown(v, t, v_\frown)$ {Take the "clockwise" branch toward t.}

24:    **end if**
25: **end while**

Note 1. The implementation of tests in lines 6, 8, 10, 12 is simple. Since the vertices of G are in convex position, one can easily (and locally—remembering only two best candidates, one for each direction) compute the first vertex in $N^+(v)$ that is in clockwise (resp. counterclockwise) direction from t or to determine that such vertex does not exist. Similarly, tests in lines 13 and 18 are simple to implement and require constant memory.

Lemma 1. *Suppose Algorithm 2 reaches a decision vertex b (line 11). Next suppose that $v_1, v_2, \ldots, v_k$ are vertices reached in subsequent steps and that all are no-choice vertices, i.e. determined at line 7 or 9. Finally suppose that next vertex reached is a dead-end vertex v_{k+1} (determined at line 12). Then vertices $v_1, v_2, \ldots, v_{k+1}$ are all either in $V_\frown(t, b, t)$ or in $V_\frown(t, b, t)$.*

Proof. By way of contradiction, suppose i is maximum so that v_i and v_{i+1} are not both in $V_\frown(t, b, t)$ or in $V_\frown(t, b, t)$. We may suppose $v_i \in V_\frown(t, b, t)$ and $v_{i+1} \in V_\frown(t, b, t)$ (The other case is analogous.), see Figure 4.

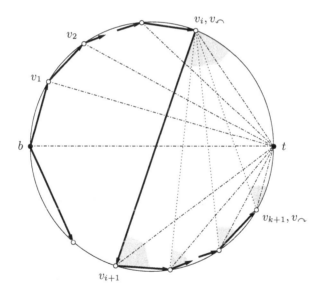

Fig. 4. Dashed parts at some vertices represent the area with no outgoing edges from the corresponding vertex in that direction. The exception are the three dotted lines which represent possible edges going into v_i. However these edges cannot help to reach t. Note that $V_\frown(b, t, b) \setminus \{t\} = V_\frown(t, b, t) \setminus \{b\}$ and $V_\frown(b, t, b) \setminus \{t\} = V_\frown(t, b, t) \setminus \{b\}$.

Since $\mathbf{G}$ is strongly connected, there must exist a path from b to t. Every such path must pass either through v_i or v_{i+1}. Since v_i is a no-choice vertex and since $v_{i+1} \in V_\frown(t, b, t)$, such a path must always pass through v_{i+1}. By the choice of i, all vertices $v_{i+1}, v_{i+2}, \ldots, v_{k+1}$ are in $V_\frown(t, b, t)$ and thus such a path must eventually pass through the vertex v_{k+1} and continue to a vertex in $N_\frown(v_{k+1}, t, v_i) \cup N_\frown(v_{k+1}, t, v_{k+1})$. However, $N_\frown(v_{k+1}, t, v_i) = \emptyset$ and $N_\frown(v_{k+1}, t, v_{k+1}) = \emptyset$, a contradiction.

Lemma 2. *Suppose Algorithm 2 reaches a dead-end vertex d (line 12). Then it will eventually return to the vertex b (last decision vertex defined in line 11).*

Proof. Suppose $v_1, v_2, \ldots, v_k$ are all vertices reached (in this order) after reaching the decision vertex b and before reaching the dead-end vertex $d = v_{k+1}$. By Lemma 1, we may assume $v_1, v_2, \ldots, v_{k+1} \in V_\frown(t, b, t)$. The other case is analogous. Since G is strongly connected, there must exist a (directed) path from d to b. Suppose by way of contradiction that the algorithm backtracks to some vertex $x \neq b$ for which $N_\frown(x, b, x) = \emptyset$. Suppose furthermore that x lies between v_i and v_{i+1} going from b to t around the outer face; see Figure 5.

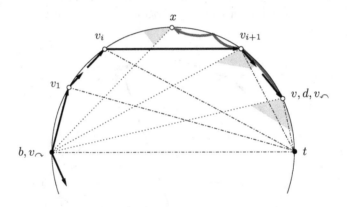

Fig. 5. Dashed parts at some vertices represent the area with no outgoing edges from the corresponding vertex in that direction. The bold curve from d to x represents the backtrack path.

By our assumption none of the edges $(v_{i+1}, b), (v_{i+2}, b), \ldots, (v_{k+1}, b)$ exist, for otherwise the backtrack procedure would follow such an edge directly to b. Hence every path from d to b must eventually pass x and continue to a vertex in $N_\frown(x, b, x)$. However $N_\frown(x, b, x) = \emptyset$ by our assumption, a contradiction.

Lemma 3. *If the "counterclockwise" branch taken at a decision vertex leads to a dead-end vertex, then no dead-end vertex is reached on the "clockwise" branch at that vertex before reaching a new decision vertex.*

Proof. The proof is similar to the two previous proofs. If $b, s_1, s_2, \ldots, s_k$ where s_k is a dead-end vertex is the counterclockwise branch at b and $b, n_1, n_2, \ldots, n_l$

where n_l is a dead-end vertex is the clockwise branch at b, then by Lemma 1, vertices $s_1, s_2, \ldots, s_k \in V_\frown(b, t, b)$ and vertices $n_1, n_2, \ldots, n_l \in V_\frown(b, t, b)$. Now it is clear that every (directed) path from b to t must pass either through s_k or n_l and then continue to a vertex either in $N_\frown(s_k, t, b) \cup N_\frown(s_k, t, b)$ or in $N_\frown(n_l, t, b) \cup N_\frown(n_l, t, b)$. However all these sets are empty by our assumption, a contradiction.

Lemma 4. *At each step of Algorithm 2 which is not the backtracking step, either $v_\frown$ or $v_\frown$ is moved closer to t (measured as the graph distance on the outer face of G.*

Proof. This follows directly from the definition of $N_\frown(v, t, v_\frown)$ and $N_\frown(v, t, v_\frown)$ and the update performed at lines 5, 7, 9, 11, and 23, respectively.

Theorem 2. *Algorithm 2 will reach t from s in at most $2n - 1$ steps.*

Proof. The proof that the algorithm will reach the destination vertex t follows from the above lemmas. The fact that no more than $2n - 1$ steps are needed follows from the fact that G has at most $2n - 1$ edges, that the algorithm process an edge at each step, and the fact that no edge is processed twice.

5   Conclusion

Routing in oriented ad-hoc networks is much more difficult than routing in non-oriented networks. Except for flooding, there seems to be no simple extension of the known routing algorithms that would be applicable to general oriented networks. In this paper we give routing algorithms for two cases of oriented planar networks: Eulerian and Outerplanar. No doubt, routing in oriented ad-hoc networks is far from being settled and further progress in this area is needed.

References

1. D. Avis and K. Fukuda. A pivoting algorithm for convex hulls and vertex enumeration of arrangements and polyhedra. In *Proc. of 7th Annu. ACM Sympos. Comput. Geom.*, pages 98–104, 1991.
2. J. A. Bondy and U. S. R. Murty. *Graph theory with applications.* American Elsevier Publishing Co., Inc., New York, 1976.
3. P. Bose and P. Morin. An improved algorithm for subdivision traversal without extra storage. *Internat. J. Comput. Geom. Appl.*, 12(4):297–308, 2002. Annual International Symposium on Algorithms and Computation (Taipei, 2000).
4. P. Bose, P. Morin, I. Stojmenovic, and J. Urrutia. Routing with guaranteed delivery in ad hoc wireless networks. *Wireless Networks*, 7:609–616, 2001.
5. E. Chavez, S. Dobrev, E. Kranakis, J. Opatrny, L. Stacho, and J. Urrutia. Traversal of a quasi-planar subdivision without using mark bits. accepted for 4th International Workshop on Algorithms for Wireless, Mobile, Ad Hoc and Sensor Networks (WMAN'04), Fanta Fe, New Mexico, 2004.

6. J. Czyczowicz, E. Kranakis, N. Santoro, and J. Urrutia. Traversal of geometric planar networks using a mobile agent with constant memory. in preparation.
7. C. Gold, U. Maydell, and J. Ramsden. Automated contour mapping using triangular element data structures and an interpolant over each irregular triangular domain. *Computer Graphic*, 11(2):170–175, 1977.
8. E. Kranakis, H. Singh, and J. Urrutia. Compass routing on geometric networks. In *Proc. of 11th Canadian Conference on Computational Geometry*, pages 51–54, August 1999.
9. F. Kuhn, R. Wattenhofer, Y. Zhang, and A. Zollinger. Geometric ad-hoc routing: Of theory and practice. In *Proc. of the 22nd ACM Symposium on the Principles of Distributed Computing (PODC)*, July 2003.
10. F. Kuhn, R. Wattenhofer, and A. Zollinger. Worst-case optimal and average-case efficient geometric ad-hoc routing. In *Proc. of the 4th ACM International Symposium on Mobile Ad Hoc Networking and Computing (MOBIHOC)*, June 2003.
11. D. Peuquet and D. Marble. Arc/info: an example of a contemporary geographic information system. In *Introductory Readings in Geographic Information Systems*, pages 90–99. Taylor & Francis, 1990.
12. D. Peuquet and D. Marble. Technical description of the dime system. In *Introductory Readings in Geographic Information Systems*, pages 100–111. Taylor & Francis, 1990.

Probabilistic Model for Energy Estimation in Wireless Sensor Networks

Mounir Achir and Laurent Ouvry

Electronics and Information Technology Laboratory
Atomic Energy Commission
17, rue des Martyrs 38054 Grenoble Cedex 09, France
{mounir.achir,laurent.ouvry}@cea.fr

Abstract. The greatest current challenge in wireless sensor networks consists in building a completely adaptive network without fixed infrastructures and with the smallest energy resources. Applications built upon such networks are various: from telemetry to medical follow-up and from intrusion detection to infrastructure maintenance. Nevertheless, their design requires a large effort to guarantee quasi unlimited lifespan as far as energy is concerned. In this work, we propose a Markov chain based model for estimating the power consumption of a wireless sensor network using CSMA/CA access protocol.

1 Introduction

Wireless sensor networks, composed of several hundreds to thousands of nodes, promise great advantages in terms of flexibility, cost, autonomy and robustness with respect to wired ones. These networks find a usage in a wide variety of applications, and particularly in remote data acquisition, such as climate monitoring, seismic activity studying, or in acoustic and medical fields. Unfortunately, the nodes are subject to strong constraints on power consumption due to their very reduced dimensions as well as their environment. As an example, frequent battery replacement is to be avoided in places where access is difficult or even impossible. Consequently, one accepts that the principal challenge remains, consequence of these constraints, the increase of the network energy efficiency in order to maximize the lifespan after deployment.

The considerable interest in wireless sensor networks led to the creation of a working group within the IEEE standardization committee, which specified a new standard IEEE 802.15.4 [1] [2]. By specifying very low power consumption MAC (Medium Access Control) and Physical layers (less than a few milliwatts), and thanks to the low complexity protocol and low data rate (no more than 250 kbits/s), this standard makes wireless sensor networks possible.

The constraint concerning low energy consumption in ad hoc networks has already led several researchers to propose protocols and algorithms, generally optimised for a precise application or certain working conditions and assumption [3] [4]. But in the case of a sensor network, context is strict since there is a single

S. Nikoletseas and J. Rolim (Eds.): ALGOSENSORS 2004, LNCS 3121, pp. 157–170, 2004.

deployment in an unreliable environment where the network must live whatever may happen.

In this article, we propose a model from which we obtain the estimate of the energy consumption in a wireless sensor node while taking into account both MAC and PHY leyers. We think that the modeling of certain states as well as the calculation of certain critical parameters permit to do this. We estimate the energy consumption with a Markovien modelisation of the CSMA/CA mechanism used in the IEEE 802.15.4 standard. The transition probabilities of the Markov chain are calculated using an interference and a traffic model described in section 5.

The remainder of this paper is organized as follows: in section 2, various approaches and solutions which we can be found in the literature are discussed. In section 3, we give a global description of our proposed model, and in sections 4 and 5, the model suggested and the dependent assumptions are described, in section 6, we present some simulation results and comments.

2   Related Works

Several solutions have been proposed in the literature to minimize energy consumption. Unfortunately each method is adapted to a particular set of assumptions and therefore can not be used to find a generic solution to optimize both quality of service and power consumption.

Concerning the MAC sublayer, one can mention for example PAPSM (Phase Announcement Power Save Mecanism) [5]. In PAPSM, each node can be in one of two states, Active or Inactive, and no synchronization between nodes is required. The principle is the following: if the duration of the active state of each node is higher than that of the inactive state, then each pair of nodes has an interval where they can communicate together. This mechanism can be implemented above any access mechanism (e.g. CSMA/CA) and is completely independent of the network layer. The inconvenient of this scheme appears when two nodes have a short communication window, this limit the amount of data which can be transmitted or increase the transmission time.

A simpler mechanism called SMAC (Synchronous Medium Access Control) was proposed in [6]. SMAC is based on the same principle as PAPSM, meaning that each station is in an active state or in an inactive state. The only difference is that the nodes must be synchronized before any data transmission and the duration of the two states is identical for all network nodes.

It is easy to see that the methods used in order to reduce consumption in the level of the MAC sublayer are always based on the principle of the radio interface extinction (Active and Inactive state). But, their strong point lies in simplicity since it is not necessary to send additional frames or to make complex calculations to reduce consumption.

3   Proposed Method

Our approach is original in that we propose a model which allows to estimate the power consumption of a node while taking into account both PHY and MAC layers. This method will be applied to the study of the IEEE 802.15.4 standard.

We propose the approach illustrated in figure 1. This approach is based on a Markov chain modelisation (figure 8) of the CSMA/CA protocol used in the IEEE 802.15.4 standard (figure 2). Starting from a set of global parameters, either defined in the standard or assumed, the transition probabilities of the Markov chain are calculated using an interference and traffic model. An estimate of the energy consumption of a network node is obtained as a consequence of the convergence of the Markov chain.

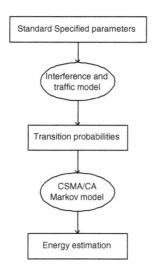

Fig. 1. Flow chart of the proposed method.

4   State Model of a Wireless Sensor Node

Wireless Local Area Networks can be classified by their medium access control protocol. One of the principal protocols used for the WLAN is the Carrier Sense Multiple Access with Collision Avoidance. According to this mechanism, all nodes wishing to transmit test the channel in order to be sure that there is not any channel activity. All packet received must be immediately acknowledged by the receiver.

4.1   CSMA/CA Mechanism

The CSMA/CA mechanism uses only one channel. Thus, only one station can successfully transmit in the network at each time. At the beginning, each station is in the Idle state. When a new packet arrives, it is stored in a buffer and the station passes to an active state ("carrier sense"). According to whether the channel is currently occupied or not, the station is then either in a backoff or transmission state.

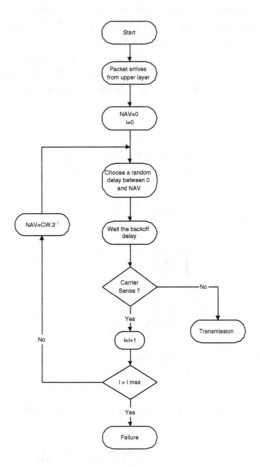

Fig. 2. Flow chart of the CSMA/CA mechanism in IEEE 802.15.4.

When the station is in a backoff state, a random duration is uniformly selected between 0 and NAV (Network Allocation Value) which represents the duration, or backoff time, the station must wait before attempting to transmit. When the backoff reaches zero, the station moves from the backoff state to the active state where a transmission is possible. If the transmission is successful, the station returns to the Idle state. It is assumed that a collision has occurred if no acknowledgement is received before the end of a predetermined time interval. A collision detection by ACK presents the advantage that it is simple and economic, and this explains its use in IEEE 802.15.4.

4.2   Markov Chain Model of the CSMA/CA Mechanism

In our model, we consider that each node can be modeled by a Markov chain [8]. Five possible state types are considered:

- **I** Idle mode;
- **A** Active (sensing, transmission attempt);
- **T** Transmission;
- **B$_i$** i^{th} backoff;
- **C$_i$** i^{th} collision.

Since there are several backoff and collision states, the total number of states is higher than five. In the Markov chain illustrated in figure 8, we calculate the transition probabilities between different states using the following probabilities:

- λ : probability of a message arrival;
- σ : probability of a message end, i.e. no more packets to transmit;
- b : probability that the channel is busy;
- f : probability of a collision;
- c : probability of a communication dropping.

Each of these probabilities is a function of network parameters such as load, node density, number of channels,... etc.

- k : probability that the backoff will not finish in the next time unit;
- s : probability that the packet transmission will not finish in the next time unit.

Here, the probability s will be constant due to the fact that we will suppose that the packets have a size which follows a defined distribution (see section 5) similar assumption will be used for the computation of k.

4.3   Power Consumption Computation

The probabilities of being in the X state, $P(X)$, are calculated with the following algorithm:

Algorithm 1

$P(I) = 1$ // Initial Condition, all other probabilities set to 0
$\pi = [P(I)\ P(A)\ P(T)\ P(B1)\ ...\ P(Bm)\ P(C1)\ ...\ P(Cm)]$
for $i = 1$ to n **do**
$\quad \pi = \pi.P$ // n very large integer, P : the transition matrix of the Markov chain
end for

Finally, the mean power consumption is calculated by the following equation:

$$Pow = P(T)Pow_T + P(I)Pow_I + P(A)Pow_A +$$
$$+ \sum_i P(B_i)Pow_B + \sum_i P(C_i)Pow_C \qquad (1)$$

with: Pow_x power consumption in the X state.

5   Probabilistic Model of a Wireless Network

In the following, we define a set of important parameters for our model:

- N: number of channels;
- R_{max}: cell radius, it can be deduced with S, P_e and the signal propagation law of the deployment environment;
- $P + 1$: node density i.e. mean number of nodes in one cell (including the desired node);
- P_e: transmission power;
- S: reception threshold i.e. receiver sensitivity;
- Ib: power level of an interferer at the receiver above which no reception is possible (assuming a given desired signal reception level equal to S);
- $AH1$: attenuated power of the adjacent channel due to the emission or reception mask;
- $AH2$: attenuated power of the alternate channel due to the emission or reception mask;
- x_E: packet duration;
- y_E: time between the start of two packets.

To calculate the distribution of the received power, P_r, we must consider an attenuation model and a node distribution model. The computation of the received power distribution is necessary to compute the probability that a communication will fail.

The calculation of this distribution uses the following assumption: nodes are distributed in a uniform way in each cell. ¿¿From this, we are able to calculate the distribution of D, the distance between a terminal and our receiver, using the following equation.

$$P_D(d) = \begin{cases} \frac{2}{R_{max}^2}d & \text{if } d \in]0, R_{max}[\\ 0 & \text{otherwise} \end{cases}$$

With this expression, one can calculate the probability density function of P_r since the received power is a function of the emitter to receiver distance and the emission power P_e, and this for a given attenuation model (antenna gains are assumed to be unitary).

We introduce a general and straightforward propagation model as follows:

$$\begin{cases} d = 10^{\frac{L_m - P_r}{10\alpha}} & \text{for } d < d_0 \\ d = d_0.10^{\frac{L_m - P_r - 10\alpha log(d_0)}{10\beta}} & \text{for } d \geq d_0 \end{cases}$$

L_m can be considered as the link margin at one meter, i.e. the transmitted power minus the loss term due to the carrier frequency. We assume the channel frequency has no impact on this term (i.e. $L_m = P_e$).

The distribution of P_r, P_{P_r}, can be calculated using the following equation:

$$P_{P_r} = \left| \frac{dd}{dp} \right| P_D(d(p)) \tag{2}$$

5.1   Traffic Model

In our study, we focus on a particular node that is trying to transmit its data. This node will be noted E and we define $P(E)$ as the probability that the node E is transmitting.

For the traffic we consider two distributions. The first one models the duration of a packet transmission, represented by the random variable x , and the second one models the duration between the start of two packets, represented by the random variable y.

Since the probability that a node is transmitting depends on these two random variables, we make the following approximation:

$$P(E) = E\left[\frac{x}{y}\right] \qquad (3)$$

Where $E[.]$ is the expected value of the random variable. $P(E)$ is also called the Duty cycle.

From this last probability one can calculate the probability that the transmissions of two nodes overlap: $P(E)P(E)$. Finally, we deduce the probability that the desired node is not transmitting while another is transmitting: $(1 - P(E))P(E)$

Now we generalize the probabilities calculated above in the case that our node is prone to several interfering sources. For that we use the elementary properties of probability theory. We also make the (strong) assumption that the properties of the traffic are identical for all network nodes.

Assuming that k' nodes in the cell have chosen node E's emission channel, the probability that k nodes among these k' nodes transmit at the same time as node E (using the Bernoulli theorem for independent events):

$$P_f^{k,k'} = P(E)C_k^{k'}(P(E))^k(1 - P(E))^{k'-k} \qquad (4)$$

This result, equation 4, will be used further on to calculate the probability of collision. In the same manner, the probability that the channel is busy is calculated using of the following equation, .i.e. the probability that k among k' nodes transmit at the same time while node E is not transmitting:

$$P_b^{k,k'} = (1 - P(E))C_k^{k'}(P(E))^k(1 - P(E))^{k'-k} \qquad (5)$$

5.2   Transition Probabilities for the Markov Chain Model

Probability that the channel is busy b: The channel is busy if at least one other transmitter in the range of our node is emitting on the same channel.

A co-channel interfering source is present if a node has selected the same channel as E and if its reception power is sufficiently high to be detected by our node using the CCA (Clear Channel Assessment) mechanism. In the following, we assume that this CCA threshold equal to Ib.

The probability that k' nodes among the P nodes in the cell choose the same channel as our transmitter is given by the following equation [9], assuming a node chooses a channel randomly among N:

$$\mu_{k'} = C_{k'}^P \left(\frac{1}{N}\right)^{k'} \left(\frac{N-1}{N}\right)^{P-k'} \tag{6}$$

Now, the probability that our receiver receives a power level higher than threshold Ib from k nodes is given by the following equation [9]:

$$\lambda_k = 1 - \left(\int_{-\infty}^{Ib} f_{P_r} dp\right)^k \tag{7}$$

Using equations 5, 6 and 7, we can compute the probability that a channel is busy. Equation 5 expresses the probability that k nodes among k' transmit at the same time and the desired node is not transmitting. Equation 6 expresses the probability that k' nodes among P choose the desired channel and finally equation 7 expresses the probability that our receiver receives a power level higher than its threshold from k nodes emitting in the desired channel. One can deduce the busy channel probability as follows:

$$b = \sum_{k=1}^{P} \lambda_k \sum_{k'=k}^{P} \mu_{k'} P_b^{k,k'} \tag{8}$$

Probability of dropping c: In order to separate the effects of co-channel interference from other types of radio interference (such as adjacent channel, image band interference, ...), transmission failures due to co-channel interference were called "collisions" (see previous section) whereas transmission failures due to these other types of interference were called drops, in our model we consider only adjacent and alternate channels effects.

For our study we suppose that this probability is only due to the interferences of the 1st and 2nd adjacent channels. Calculation of this probability is based on [9] and is given by the following equation, for the adjacent channels:

$$c_1 = \sum_{k=1}^{P} \lambda_k' \sum_{k'=k}^{P} P_f^{k,k'} \left(\frac{2}{N}\mu_{k'(edge)} + \frac{N-2}{N}\mu_{k'(center)}\right) \tag{9}$$

with

$$\mu_{k'(edge)} = C_{k'}^P \left(\frac{1}{N}\right)^{k'} \left(\frac{N-1}{N}\right)^{P-k'} \tag{10}$$

$$\mu_{k'(center)} = C_{k'}^P \left(\frac{2}{N}\right)^{k'} \left(\frac{N-2}{N}\right)^{P-k'} \tag{11}$$

$$\lambda_k' = \left(1 - \left(1 - \int_{Ib+AH1}^{P_e} f(P_r) dp\right)^k\right) \tag{12}$$

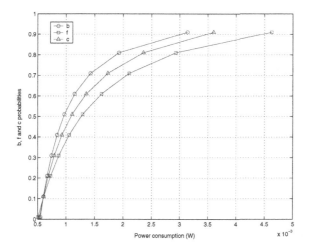

Fig. 3. b, f and c in function of power consumption.

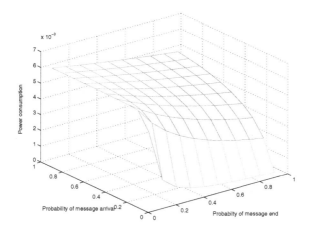

Fig. 4. Power consumption in function of λ and σ.

For the alternate channels one has:

$$c_2 = \sum_{k=1}^{P} \lambda_k'' \sum_{k'=k}^{P} P_f^{k,k'} \left(\frac{4}{N} \mu_{k'(edge)} + \frac{N-4}{N} \mu_{k'(center)} \right) \tag{13}$$

$$\lambda_k'' = \left(1 - \left(1 - \int_{Ib+AH2}^{P_e} f(P_r)dp \right)^k \right) \tag{14}$$

The reader can notice that in this formula one must use $AH1$ and $AH2$ to take into account emitter and receiver attenuation [9] masks as specified in the standard.

For the computing of c, one uses the following equation:

$$c = c_1 + c_2 - c_1 c_2 \tag{15}$$

k and s Probabilities: In the CSMA/CA mechanism, the backoff duration is selected in a uniform manner in an interval which varies according to the number of attempts. Noting T_{ave} the average duration of a backoff, one can approximate k with:

$$k = P_{backoff}\left(t \leq T_{ave}\right) = F_{backoff}\left(T_{ave}\right) \tag{16}$$

Since the distribution of the Backoff time is a uniform distribution, it is easy to deduce that $k = 0.5$.

One applies the same approximation to find the probability that the transmission of a packet will not finish in the next time unit but this time T_{ave} represents the average duration of a packet transmission.

$$s = P_{Transmission}\left(t \leq T_{ave}\right) = F_{Transmission}\left(T_{ave}\right) \tag{17}$$

In [10], we suppose that the size of packets follows an exponential distribution whose expression is as follows:

$$P_\mu(t) = \mu e^{-\mu t} \implies F_\mu(t) = 1 - e^{-\mu t} \tag{18}$$

The mean is given by: $E[t] = 1/\mu$, therefore:

$$s = F_\mu(E[t]) = 1 - e^{-1} = 0.6321 \tag{19}$$

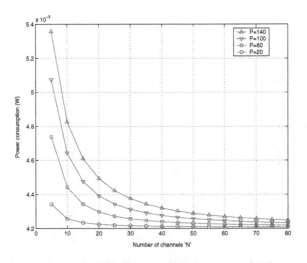

Fig. 5. Power consumption as a function of N.

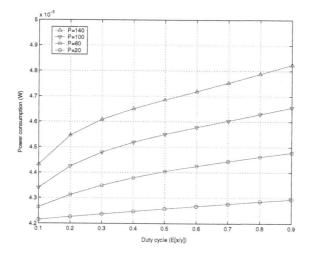

Fig. 6. Power consumption as a function of the duty cycle ($E[x/y]$).

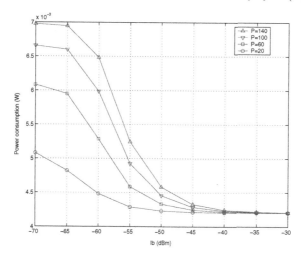

Fig. 7. Power consumption as a function of Ib (dBm).

6   Results

Our objective is to see the impact of the considered parameters, in our model, on energy consumption, and this while taking account the effect of the MAC sublayer. i.e. with the mechanism used for medium access. The following curve shows the probabilities b, f and c as functions of the energy consumption, computed with the algorithm see in section 4. Each curve is plotted assuming the other too probabilities equal to 0.1.

We see that, compared to the values that we took, the influence of the collision probability on energy consumption is more significant than that of dropping a communication or detecting an occupied channel (see figure 3).

Table 1. Power consumption for different node states.

Node state	Power consumption (mW)
Idle	10^{-2}
Actif	1
Transmission	10
Backoff	1
Collision	10

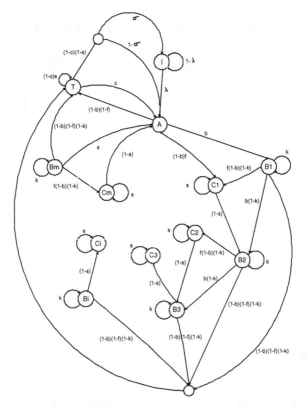

Fig. 8. Markov chain of the CSMA/CA mechanism.

Figure 4 shows the variation of power consumption as a function of the probability of message arrival λ and the probability of message end σ. Obviously if there is a small message end probability and a significant probability of message arrival, one obtain a maximum power consumption, as confirmed by figure 4.

In the following, we will see the results obtained when we vary the number of channels, the interference threshold and the duty cycle. The results are obtained with the following values (similar to the IEEE 802.15.4 specifications):

$\lambda=0.25$, $\sigma=0.5$, $N=16$, $P=20$, 60, 100 and 140, $Ib=-70$ dBm, $P_e=0$ dBm, $AH1=20$ dBm, $AH2=20$ dBm, $S=-85$ dBm, $\alpha=2$, $\beta=3.3$ [2].

Figure 5 shows the node power consumption as a function of the number of channels, which obviously decreases as N increase owing to the fact that one will have less collisions and less communications drops since the probability of co-channel and adjacent channel interference will decrease.

Figure 6 shows the variation of the power consumption as a function of the duty cycle and one can notice that it is nearly a linear variation.

In figure 7, one sees well that, for great densities (like here for $P=140$), one has a gain in consumption of approximately 20% if one increase the interference threshold by 5 dBm (from -60 dBm). This gain decrease considerably for small densities which is obvious since one will have much less interference.

7   Conclusion

In this article, we have presented a model where each network node is modeled by a Markov chain. We showed how, starting from this model, to calculate the power consumption of the nodes. We model also the traffic, interferences and the channel chosen to compute each one of the probabilities used in the Markov model.

The synthesis of this work gives us a relation ship between the nodes power consumption and each of the main parameters specified in the physical layer and this taking into account the effect of the mechanism used for medium access control (CSMA/CA).

Future work will consist in testing, for each type of application type, the performance in term of power consumption in order to provide the optimal configuration for each application.

References

1. Ed Callaway, Venkat Bahl, Paul Gorday, Jose A. Gutierrez, Lance Hester, Marco Naeve and Bob Heile, "Home Networking with IEEE 802.15.4: A Developing Standard for Low-Rate Wireless Personal Area Networks, "IEEE Communications MAGAZINE, Special issue on Home Networking, August 2002, pp. 70-77.
2. Draft P802.15.4/D18, Fevrier-2003: Wireless Medium Access Control (MAC) and Physical Layer (PHY) specifications for Low Rate Wireless Personel Area Networks (LR-WPANs).
3. Rahul C.Shah and Jan M.Rabaey, Energy Aware Routing for Low Energy Ad hoc Sensor Networks 2002.

4. Wendi Rabiner Heinzelman, Anantha Chandrakasan and Hari Balakrishnam, Energy Efficient Communication Protocol for Wireless Microsensor Networks (LEACH) 2000.

5. Laura Marie Feeney, A QOS Aware Power Save Protocol for Wireless Ad hoc Networks, 3rd Swedish Networking Workshop, Marholmen, September 11-14, 2002.

6. Wei Ye, John Heidemann, and Deborah Estrin. An Energy-Efficient MAC protocol for Wireless Sensor Networks. In Proceedings of the IEEE Infocom, pp. 1567-1576. New York, NY, USA, USC/Information Sciences Institute, IEEE. June, 2002.

7. Tamer M.S. Khattab, Mahmoud T. El-Hadidi and Hebat-Allah M. Mourad; Analysis of Wireless CSMA/CA Network Using Single Station Superposition (SSS); International Journal of Electronics and Communications 2002 N°2, 71-81.

8. Raquel A. F. Mini, Badri Nath, Antonio A. F. Loureiro; A Probabilistic Apprach to Predict the Energy Consumption in Wireless Sensor Networks; IV Workshop on Wireless Communication, São Paulo, Brazil, October 23-25 2002.

9. C.Bernier, P. Senn; A Probabilistic Method for Designing RF Specifications of High Density Ad Hoc Networks; World Telecommunication Congress, 22-27 September 2002.

10. Draft P802.15.2/D01, 2001: Recommanded Practice for Telecommunications and Information exchange between systems -LAN/WAN- Specific Requirements -Part 15: Recommanded Pratcice for Wireless Personal Area Networks Operating in Unlicensed Frequency Bands.

Multi: A Hybrid Adaptive Dissemination Protocol for Wireless Sensor Networks[*]

Carlos M.S. Figueiredo[1,2], Eduardo F. Nakamura[1,2], and
Antonio A.F. Loureiro[1]

[1] Department of Computer Science
Federal University of Minas Gerais
Belo Horizonte, MG, Brazil
{mauricio, nakamura, loureiro}@dcc.ufmg.br
[2] FUCAPI – Analysis, Research and Technological Innovation Center
Manaus, AM, Brazil
{mauricio.figueiredo, eduardo.nakamura}@fucapi.br

Abstract. Data dissemination (routing) is a basic function in wireless sensor networks. Dissemination algorithms for those networks depend on the characteristics of the applications and, consequently, there is no self-contained algorithm appropriate for every case. However, there are scenarios where the behavior of the network may vary a lot, such as an event-driven application, favoring different algorithms at different instants. Thus, this work proposes a new hybrid and adaptive algorithm for data dissemination, called Multi, that adapts its behavior autonomously in response to the variation of the network conditions. Multi is based on two algorithms for data dissemination that are also presented and evaluated in this work: SID (Source-Initiated Dissemination), a reactive algorithm where dissemination is started by the source nodes, and EF-Tree (Earliest-First Tree), an algorithm that builds and maintains a tree, in a proactive fashion, to disseminate data towards the sink.

1 Introduction

A wireless sensor network (WSN) [1,2] is comprised of a large number of sensor nodes forming an *ad hoc* network to monitor an area of interest. This type of network has become popular in the scientific community due to its applicability that includes several areas, such as environmental, medical, industrial, and military.

WSNs diverge from traditional networks in many aspects. Usually, these networks have a large number of nodes that have strong constraints like power restrictions and limited computational capacity. In general, WSNs demand self-organizing features, i.e., the ability of autonomously adapt to the eventual changes resulted from external interventions, such as topological changes (due

---

[*] This work was partially supported by CNPq, Brazilian Research Council, under process 55 2111/02-3.

S. Nikoletseas and J. Rolim (Eds.): ALGOSENSORS 2004, LNCS 3121, pp. 171–186, 2004.

to failures, mobility or node inclusion), reaction to an event detected, or due to some request performed by an external entity.

The objective of such network is to collect data from the environment and send it to be processed and evaluated by an external entity through a sink node. Consequently, data dissemination towards the sink node is a fundamental task and, depending on the application, it can be performed considering different models [3]. In a continuous monitoring, a monitoring application receives continuously sensing data from the environment. In an event-driven monitoring, a sensor node sends a notification to the monitoring node when an event of interest happens. In an observer-initiated monitoring, an observer sends a request to the network that gathers data about the sensing data and sends it back to the monitoring node. In a hybrid strategy, a WSN can also allow different types of data delivery, like in a scenario that reports the gathered data whenever the observer requests it, and the sensor node immediately sends a report of a critical event whenever it happens.

Different algorithms [4,5,6,7] have been proposed to disseminate (routing) data gathered by sensors. However, different applications and scenarios demand algorithms with different features. This will be shown in this work through the presentation and analysis of two algorithms: SID (Source-Initiated Dissemination), a reactive algorithm for event driven scenarios; and EF-Tree (Earlier-First Tree), a proactive algorithm for scenarios with intense data communication.

Thus, given a specific scenario, the WSN can be designed to operate with the most appropriated data dissemination algorithm, which can be defined *a priori*. However, in some cases the variations of these scenarios can be constant or even unpredictable. For instance, an event-driven scenario may present a low incidence of events and, at a given moment several events can be detected generating a high traffic. For those situations, it should be provided algorithms, which are appropriated for different periods, such as SID and EF-Tree, and it might be infeasible or undesirable to an external entity to act dynamically on the network to change its behavior.

Ideally, the network should be able to self-organize itself according to such variations, adjusting its behavior to assure that all functions (e.g., sensing, communication and collaboration) will be energy-efficient during its lifetime. This work presents and evaluates a hybrid algorithm that adapts itself in a scenario with variations of the traffic load. The algorithm, called Multi, aggregates the features of both, SID and EF-Tree, and autonomously alternates its operation to fit the current condition. This approach represents a new strategy to build data dissemination algorithms for WSNs.

The remaining of this paper is organized as follows. Sections 2 and 3 describe the data dissemination algorithms EF-Tree and SID, respectively. Section 4 presents Multi, an adaptive algorithm based on EF-Tree and SID. The evaluation of the algorithms is presented in Section 5. Section 6 compares our solution to other algorithms presented in the literature. Finally, Section 7 presents our final considerations and discusses some future directions.

2   EF-Tree (Earlier-Firt Tree)

Usually, WSNs collect data from the environment delivering it to a more powerful system that stores, analyzes and deliveries such data to the users. Generally, there is an special element, called sink node, responsible for connecting the WSN to an infra-structured environment. Thus, data collected by the sensor nodes must be disseminated towards a sink node, usually in a multi-hop fashion.

A simple and efficient structure for data dissemination is a routing tree [8]. This structure is created and maintained by a sink node in a proactive fashion, connecting all reachable nodes. Our implementation, called EF-Tree (Earlier-First Tree), has the particularity of rebuilding periodically the tree to cope with eventual topological changes. The building process works as follows:

- The sink node starts the process by broadcasting a control message as depicted in Fig. 1(a).
- When a node initially receives the building message, it identifies the sender as its parent and broadcasts the building message. Messages received from other neighbors are discarded. Note that it is possible to define other possibilities to choose a parent node such as to choose the node that belongs to a path with the highest amount of energy available.
- Whenever a node has a data to be transmitted (sensed or forwarded by another node), it will send it directly to its parent (Fig. 1(b)).
- The building process is periodically repeated so the network reflects eventual topological changes, such as failures, node movements and inclusion of new nodes. This periodicity depends on the frequency topological chances occur. More dynamic networks need shorter rebuilding periods.

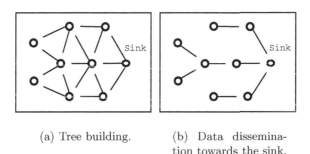

(a) Tree building.   (b) Data dissemina-
                          tion towards the sink.

Fig. 1. EF-Tree algorithm.

3   SID (Source-Initiated Dissemination)

SID is a reactive algorithm for data dissemination. In this case, the process of creating the data dissemination infra-structure is initiated by the source node whenever necessary (e.g., when an event is detected). This principle is appropriate for applications that disseminate data whenever an event is detected.

For event-driven scenarios, this approach is better than a proactive algorithm, such as EF-Tree, because it does not need to continuously maintain an infrastructure for data dissemination. This is particularly true in scenarios where events have a low occurrence rate (hours or days).

In this approach we assume that sensors are already deployed and preconfigured to detect an event of interest, in the same way a sprinkler is already configured to detect fire. The network will remain inactive until events are detected and the communication process is started by sensors that detected them. The algorithm works as follows:

- Nodes that detect an event broadcast their data, the node's identification and a timestamp (Fig. 2(a)). This allows the identification of the data.
- Whenever a node receives the data sent by another node, it stores its identification (source identification and timestamp), as well as the sender identification. As the disseminated data is broadcasted, the node will receive it from all neighbors, however, it will register and forward only once (the first data received). Like the EF-Tree, another criteria can be considered.
- Similarly, data will arrive at the sink node from all of its neighbors. Then, the sink will send a control message requesting the data to be sent by the node from which it firstly received the data. This message identifies the data to be sent.
- When a node receives this control message, it repeats this process identifying which node should send that data to it. This process is repeated until the source nodes are reached (Fig. 2(b)).
- Once a source node receives the control message requesting its data, it will update its own table so subsequent data will be sent to the node that firstly requested its data. Thus, data will be disseminated through the path where the sink's requisition message arrived, which might be the fastest path (Fig. 2(c)).
- In order to allow the network to adjust to eventual topological changes (due to failures, mobility or node inclusion), the requisition messages are periodically sent by the sink towards the sources while data is being received. Once a node (source or intermediate) stops receiving requisition messages, due to any topological change, the node will restart to send or forward data in broadcasts. Thus, if any path exists, data will reach the sink again and it will restart the requisition process.
- Once the events disappear, data will not be generated anymore and, consequently, the sink node will stop sending requisition messages to the sources. In absence of the periodical requisition messages of a specific data, the table entries will expire and the network will become inactive again.

The idea of requesting data periodically, finding paths between sources and a sink node is interesting mainly when an event occurs and does not stop immediately. The frequency of requesting data should be lower than the data generation

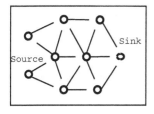

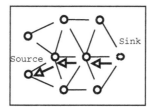

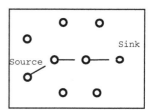

(a) Data diffusion.

(b) Requisition mes-
sages.

(c) Data dissemina-
tion towards the sink.

Fig. 2. SID algorithm.

avoiding the overhead of flooding every data and the overhead of sending a requisition message for each data.

4 Multi: A Hybrid Adaptive Protocol for Data Dissemination

Clearly, in the algorithms described above (SID and EF-Tree), we can observe
that they are appropriate for different scenarios. Comparatively, SID should
used whenever the event ratio is low, so there is no need to constantly maintain
a network infra-structure such as EF-Tree does. With SID the infra-structure
is created and maintained on demand, i.e., only when there is any data to be
transmitted.

When the communication load is high, such as in scenarios with continuous
sensing and many sources, SID is not an appropriate choice because it maintains
a dissemination scheme for each source, which is started by a flooding causing a waste of energy and bandwidth. In such cases, EF-Tree is clearly more
appropriate.

As previously mentioned, in an event-driven scenario, the events might occur
with a non-uniform distribution. For instance, the network may remain with a
very low activity for days, which is good for SID, however, in a given moment
several events might occur generating a traffic large enough to use the EF-Tree.
Choosing an algorithm a priori surely will not achieve the best results during the
entire network lifetime. Thus, we propose a hybrid and adaptive algorithm, called
Multi, that incorporates the features of both algorithms adapting its behavior
according to the current network condition. This approach, to the best of our
knowledge, is new in the WSN domain. Furthermore, in this work we propose a
new class of algorithms for self-organization.

The current implementation of Multi incorporates the characteristics of SID
and EF-Tree. The adaptive control of Multi will be performed by the sink node
that monitors the amount of events detected by the network, since both EF-Tree

and SID control schemes are performed by the sink node. The operation of Multi obeys the following steps:

- Initially, the network operates just as SID. When a node detects an event it floods the data in the network. Once the sink receives the data, it sends requisition messages towards the sources, so the paths between each source and the sink is formed. These requisitions are sent periodically while the sources keep sending data.
- The sink node computes the amount of sources sending data in a predefined period. In our implementation, we made this period to be the same of the requisition period.
- Once the amount of sources exceeds a given threshold, the sink node begins to send messages to build the tree, just as the EF-Tree does. This comes from the observation that for a given number of sources, we can wait its increase, and it is less costly to build and maintain a dissemination infra-structure to the entire network than for every possible source.
- When a node receives a building tree message, it keeps the parent's information for a validation time, which is greater than or equal to the periodicity of the building messages. After this, if the node has a valid parent it will always send its data to the current parent.
- If the event rate diminishes to a value lower than the predefined threshold, the sink node will stop building the tree. Consequently, the parent of each node will become invalid (the validation period will expire) so the network will operate again as SID.

The definition of the threshold value for changing the routing algorithm depends on several factors that impact the execution cost of each algorithm. For instance, the network size, the amount of sources (traffic), traffic duration and data rate. Thus, this threshold should be dynamically computed based on models of event occurrence and the previous behavior. However, for the sake of simplicity, it will be defined a static threshold based on the simulations of SID and EF-Tree algorithms.

5   Simulation and Evaluation

In this section, the algorithms previously described are evaluated through simulation. Initially, we will describe some particular scenarios assessing the performance of SID, EF-Tree, Flooding and Directed Diffusion. Afterwards, we will show the adaptive ability of Multi and its advantages compared to SID and EF-Tree alone.

The experiments were performed using the ns-2 Network Simulator [9]. The simulation parameters were based on the Mica2 Sensor Node [10] using the 802.11 protocol in the MAC layer (see Table 1). In all simulations, we considered only one sink, data messages of 20 bytes were transmitted every 10s, and control messages of 16 bytes were transmitted every 100s.

Table 1. Mica2 parameters used in the simulations.

Parameter	Value
Transmission power	33,3mW
Reception power	30,0mW
Bandwidth	76800 bps
Communication radius	40m

5.1 SID Versus EF-Tree

To compare the behavior of SID and EF-Tree we defined two scenarios. In the first one, data is disseminated continuously, where several nodes send their data periodically towards the sink. In this scenario we will evaluate the scalability of the algorithms according to the size of the network, and their resilience in presence of failures. In the second scenario, events occur randomly with random duration. The metrics evaluated were the delivery ratio to assess failures, and energy consumption, the most restrictive resource in a WSN.

Both algorithms are also compared to both the classical flooding approach and the Directed Diffusion algorithm. For the Directed Diffusion, interests were disseminated every 100s (equal to the building period of EF-Tree and data request in SID), interests were valid for 200s and negative reinforcement was enabled.

Figure 3 shows the behavior of the algorithms in a scenario with 50 nodes distributed uniformly over a $100 \times 100m^2$ area, with continuous traffic. The number of sources varied from 2 to 50 nodes, randomly chosen.

As we can see in Fig. 3, flooding a data impacts the delivery rate (Fig. 3(a)) on a network with bandwidth constraints. This can be observed in the Flooding algorithm with the decrease of delivered packets when the number of sources increases. The Directed Diffusion, which floods interests and data to setup gradients, we can observe a more drastic impact given by the loss of control messages, because paths are not reinforced so broadcasts are maintained. This occurs because the network cannot process the data rapidly, overfilling the queues of routing nodes, which results in losses. SID also starts the dissemination process by flooding the data, however, once a path connecting sources to a sink are established, no data is flooded anymore and the queues of routing nodes are not overflowed, keeping high delivery rates such as EF-Tree.

Regarding the energy consumption, EF-Tree presented the best result (Fig. 3(b)), showing the advantage to build proactively the dissemination infrastructure to the network in this scenario. SID has a higher increase on energy consumption because of the need to build this infra-structure for every source. In addition, Directed Diffusion consumes more energy than EF-Tree and SID, and less than Flooding with two sources, and operates saturated with more than ten sources. This poor performance is consequence of the impact of the delivery rate, as described before, and shows that it is not viable for the simulated scenario (low bandwidth and high number of sources).

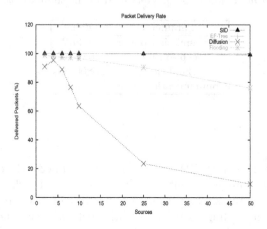

(a) Packet delivery ratio.

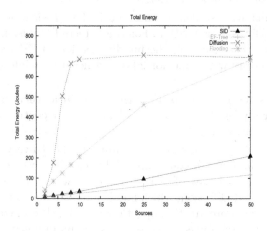

(b) Absolute energy consumption.

Fig. 3. Continuous data traffic in a 50-node network.

The scalability of the algorithms was evaluated in the previous scenario, however, the number of sources was fixed (20 nodes) and the network size varied from 50 to 200 nodes. As we can see in Fig. 4, the impact of the network size is greater in the algorithms based on flooding, because of packet losses, as described before (Fig. 4(a)). Clearly, in this scenario, EF-Tree also scales better than the others when we evaluate the energy consumption (Fig. 4(b)). SID spent more energy than EF-Tree due to the initial flooding of data and, again, Directed Diffusion presented a worse behavior than SID and EF-Tree, working saturated.

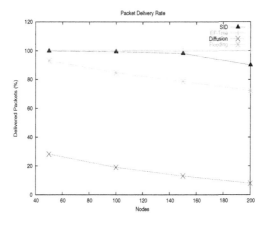

(a) Packet delivery ratio.

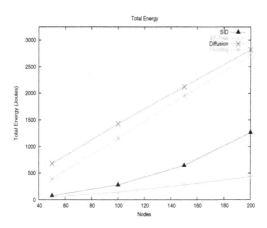

(b) Absolute energy consumption.

Fig. 4. Scalability.

To evaluate the resilience of the algorithms, we fixed the number of sources in 20 nodes and varied the probability of failures from 0 to 50% randomly during the simulation. When a node fails, it stays inactive until the simulation ends. As we can see in Fig. 5, SID and EF-Tree outperform the others when the failure probability is low. Their delivery rate decreases and stays close to 90% with the increase of failures. Flooding maintains its performance close to 90%, with losses related to high broadcasts. The Directed Diffusion has low delivery rate due to the saturated operation as in the previous simulations.

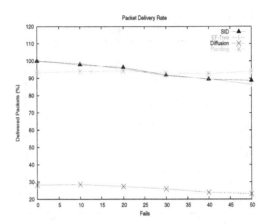

Fig. 5. Packet delivery ration under different probabilities of failure.

Now we turn to the second scenario, based on the previous one that represents an event-driven application. To represent the random occurrence of events, sources generated data randomly along the simulation time. The duration of data generation was a random value between 1 and 50s.

As we can see in Fig. 6(b), SID outperforms the other algorithms when the number of sources is low. This occurs because the dissemination infra-structure created by SID tries to use only the necessary nodes connecting the sources to the sink node. Besides, this infra-structure is created and maintained only when necessary. However, when the traffic increases, EF-Tree begins to be more adequate because it builds the dissemination infra-structure for all nodes at once. We can also observe that when the traffic is not very intense, all algorithms present better packet delivery ratio (Fig. 6(a)). Comparing to the previous scenario, Directed Diffusion improved its performance, which reinforces the fact that it is appropriate to scenarios with few sources only.

5.2 MULTI

To illustrate the adaptability of Multi, we created a scenario where the occurrence of events varies along the simulation time. Basically, we generated a wave of events in the network, as it occurs when a mobile target is detected. At the instant the target begins to invade the sensor field few nodes are able to detect its presence. When the target is fully present at the sensor field the number of sensors that detects it is maximized. When it begins to leave the sensor filed, again, few nodes report its presence, and when the target is gone, the network becomes inactive again. We configured Multi to change the routing strategy when the threshold of three sources is identified. This value was chosen based on the previous simulations that showed that SID outperforms EF-Tree for scenarios of few sources (less than six sources).

In Fig. 7 each point of the curves represents the energy consumption accumulated in the last 10 seconds. In these simulations, data was generated from

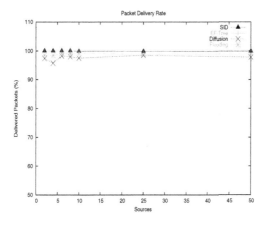

(a) Packet delivery ratio.

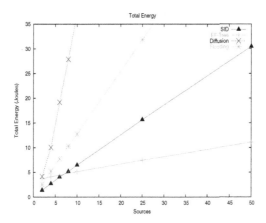

(b) Absolute energy comsumption (Ampliated).

Fig. 6. Random events of random duration.

the instant 500 to 900s. Fig. 7(a) shows that while the number of sources does not reach the threshold, Multi behaves exactly as SID outperforming EF-Tree in the intervals of inactivity (0 to 500s and 900 to 1400s), because EF-Tree keeps building the tree even when no data is generated (energy increases at every 100s). At the beginning of the activity period (about 500s) the consumption of SID and Multi started to increase because of the initial data flooding started by the sources. Shortly after, Multi consumption is similar to EF-Tree because the paths connecting sources to sink are established.

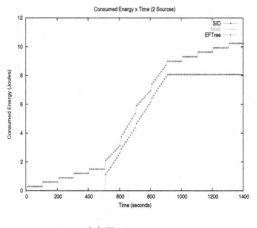

(a) Two sources.

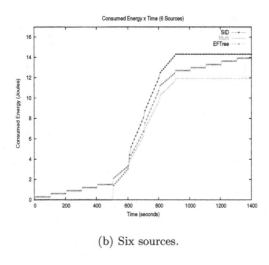

(b) Six sources.

Fig. 7. Traffic based on a wave of events.

When the number of sources increases, we noticed that SID consumed more energy compared to EF-Tree during the period of data traffic (Fig. 7(b)). This occurs because of the larger number of nodes flooding its data (instants 500s and 600s in Fig. 7(b)). At this point, Multi adapts its behavior as soon as some sources begin to generate data starting to operate as EF-Tree and outperforming SID. When the traffic reduces again (instants 800 and 900s in Fig. 7(b)), Multi changes it behavior again to operate like SID. This adaptation allows to save energy compared to SID and EF-Tree.

We also evaluated the behavior of Multi under the scenarios of exhaustive simulation previously used to evaluate SID and EF-Tree. The energy consumption of each algorithm at the two different scenarios is depicted in Fig. 8.

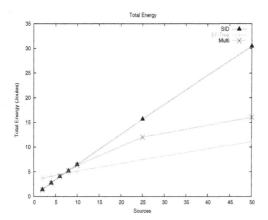

(a) Energy consumption in the event-based scenario.

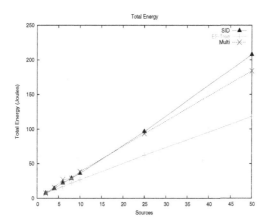

(b) Energy consumption in the continuous scenario.

Fig. 8. Multi versus SID versus EF-Tree.

The results showed that Multi outperforms EF-Tree when the number of sources is low because in such cases the algorithm behaves as SID. When the number of sources increases, Multi starts to behave as EF-Tree outperforming

SID, which is not suitable for such scenario (Fig. 8(a)). This happens because the simulation time was kept constant and the increase in the number of sources leads to a decrease in the inactivity time amortizing the cost of the EF-Tree. In a real event-driven scenario, we can expect to have longer inactivity periods than in the simulated scenarios. Thus, the energy cost of the EF-Tree is shifted to a higher value due to its proactive characteristic what results in a better performance of both SID and Multi.

After the transition points, we can observe that the energy cost of Multi is not the same of EF-Tree that happens because Multi always starts to operate as SID changing to EF-Tree only when the traffic threshold is reached. The difference observed at the transition points depends on the conditions of the simulated scenarios (simulation time, data rate and network size). In the scenario of continuous traffic (Fig. 8(b)), Multi does not present great advantages compared to SID because all traffic was generated in the beginning of the simulation, so Multi adapted only when the greater communication cost had already occurred. An advantage can be observed in Multi, though, compared to SID when the number of sources is more than 25, and it happens because the energy cost of building a tree is less than the cost of sending individual messages to a large number of sources.

In an event-driven scenario, it is worth to mention that the more sporadic the event is, the greater the advantages of using Multi rather than SID or EF-Tree. On the other hand, in a scenario of intense traffic, it might be preferable to use only EF-Tree.

6   Related Work

Several routing algorithms have been proposed to solve the data dissemination problem in WSNs [4,5,6,7]. However, none of these algorithms presents a definitive solution for different scenarios. In fact each algorithm is designed do operate in a specific scenario and application. Directed Diffusion [4] was the first algorithm for data-centric routing in WSNs and, although, it is able to adapt to eventual topological modifications it is not the most appropriate solution for event-driven scenarios because it needs to flood the interests periodically, which results in waste of resources (specially when the number of sources is large). There are few data dissemination algorithms designed for event-driven scenarios. An example is SPIN [7] that tries to suppress the weakness of flooding by local negotiations (that use data identifications only) so data is sent only for nodes that are interested in them.

Recently, it was proposed another work, related to the one presented in this work, that deals with the necessity for matching the data dissemination algorithms to the application characteristic [11]. In that work, two algorithms, extended from Directed Diffusion, are proposed: Push Diffusion, for few sources; and One-Phase Pull Diffusion, for a high number of sources. The characteristics of those algorithms are similar to SID and EF-Tree. However, this work consists of a new approach to design data dissemination algorithms exemplified by

Multi, which incorporates the features of two other algorithms adapting itself automatically when certain communication conditions are verified. Also, a more realistic evaluation was done by the simulation with real sensors network parameters (based on Mica2), while the work in [11] used an ideal scenario counting only the number of transmitted messages.

Other efforts related to the design of adaptive and hybrid algorithms can be found in the literature. For example, SHARP [12] is a routing protocol for ad hoc networks that founds an equilibrium between reactive and proactive protocols adapting the degree of how routing information is propagated in the network. This work is different by focusing on data dissemination characteristics of WSNs.

7   Conclusions and Future Work

In this work we presented a new approach for self-organizing wireless sensor networks to allow the autonomous adaptation of the data dissemination strategy to improve the energy efficiency when the scenario presents different variations on its traffic condition. This was done through the presentation and evaluation of two different algorithms for data dissemination in WSNs, SID and EF-Tree, showing their behavior in different scenarios, and introducing Multi, that unites the behavior of both algorithms so it can adapt itself according to the number of events that are detected by the sensor nodes.

It is true that in WSNs each algorithm is appropriate for a different application and a different scenario. However, in scenarios that present high variability on the incidence of events (or data traffic), only one algorithm might not achieve the best result along the network lifetime. When operational conditions of the network are unpredictable, adaptive hybrids algorithms like Multi should be more adequate than a single strategy-algorithm, extending its applicability.

Although the current implementation and evaluation of Multi did not explored its full potential, it clearly provides benefits in some situations, such as the the ones illustrated here. Thus, to make it more generic, its adaptive capability should depend on event occurrence models and prediction models based on the previous behavior.

Future work includes the evolution of Multi, introducing data aggregation functions and adoption of different behaviors for different partitions of the network, according to the model of event occurrence. Also, we plan to design new hybrid adaptive algorithms based on the approach presented here, however, including other data dissemination algorithms such as Push Diffusion and One-Phase Pull Diffusion briefly mentioned above.

References

1. Pottie, G.J., Kaiser, W.J.: Wireless integrated network sensors. Communications of the ACM **43** (2000) 51–58
2. Akyildiz, I.F., Su, W., Sankarasubramaniam, Y., Cyirci, E.: Wireless sensor networks: A survey. Computer Networks **38** (2002) 393–422

3. Tilak, S., Abu-Ghazaleh, N.B., Heinzelman, W.: A taxonomy of wireless micro-sensor network models. ACM Mobile Computing and Communications Review (MC2R) **6** (2002) 28–36

4. Intanagonwiwat, C., Govindan, R., Estrin, D.: Directed diffusion: A scalable and robust communication paradigm for sensor networks. In: Proceedings of the 6th ACM International Conference on Mobile Computing and Networking (Mobi-Com'00), Boston, MA, USA, ACM Press (2000) 56–67

5. Heinzelman, W., Chandrakasan, A., Balakrishnan, H.: Energy-efficient communi-cation protocols for wireless microsensor networks. In: Proceedings of the Hawaiian International Conference on Systems Science (HICSS), Maui, Hawaii, USA (2000)

6. Ganesan, D., Govindan, R., Shenker, S., Estrin, D.: Highly-resilient, energy-efficient multipath routing in wireless sensor networks. ACM SIGMOBILE Mobile Computing and Communications Review **5** (2001) 11–25

7. Kulik, J., Heinzelman, W., Balakrishnan, H.: Negotiation-based protocols for dis-seminating information in wireless sensor networks. Wireless Networks **8** (2002) 169–185

8. Sohrabi, K., Gao, J., Ailawadhi, V., Pottie, G.: Protocols for self-organization of a wireless sensor network. IEEE Personal Communications **7** (2000) 16–27

9. NS-2: The network simulator - ns-2 (2004) [Online] Available: http://www.isi.edu/nsnam/ns/.

10. Crossbow: Mica2 - wireless measurement system (2004) [Online] Available: http://www.xbow.com/Products/New_product_overview.htm.

11. Heidemann, J., Silva, F., Estrin, D.: Matching data dissemination algorithms to application requirements. First ACM Conference on Embedded Networked Sensor Systems (SenSys 2003) (2003)

12. Ramasubramanian, V., Haas, Z., Sirer, E.: SHARP: A hybrid adaptive routing protocol for mobile ad hoc networks. In: Proceedings of the 4th ACM Interational Symposium on Mobile Ad Hoc Networking and Computing (MobiHoc '03). (2003) 303–314

Constrained Flow Optimization with Applications to Data Gathering in Sensor Networks[*]

Bo Hong and Viktor K. Prasanna

University of Southern California, Los Angeles CA 90089-2562
{bohong, prasanna}@usc.edu

Abstract. We focus on data gathering problems in energy-constrained wireless sensor networks. We study store-and-gather problems where data are locally stored on the sensors before the data gathering starts, and continuous sensing and gathering problems that model time critical applications. We show that these problems reduce to maximization of network flow under vertex capacity constraint, which reduces to a standard network flow problem. We develop a distributed and adaptive algorithm to optimize data gathering. This algorithm leads to a simple protocol that coordinates the sensor nodes in the system. Our approach provides a unified framework to study a variety of data gathering problems in sensor networks. The efficiency of the proposed method is illustrated through simulations.

1 Introduction

A major application of the sensor networks is to periodically monitor the environment, such as vehicle tracking and classification in the battle field, patient health monitoring, pollution detection, etc. The basic operation in such applications is to sense the environment and eventually transmit the sensed data to the base station for further processing. The data gathering processes are usually implemented using multi-hop packet transmissions, since short range communications can save energy and also reduce communication interferences in high density WSNs [1].

Although the applications may be designed for different functionalities, the underlying data gathering processes share a common characteristic: compared with sensing and computing, transferring data between the sensors is the most expensive (in terms of energy consumption) operation [1]. In this paper, we study the energy efficiency of data gathering in the WSNs from an algorithmic perspective. By modeling the energy consumption associated with each send and receive operation, we formulate the data gathering problem as a constrained network flow optimization problem where each each vertex u is associated with a capacity constraint w_u, so that the total amount of flow going through u (incoming plus outgoing flow) does not exceed w_u. We show that such a formulation models a variety of data gathering problems (with energy constraint on the sensor nodes).

Some variations of the data gathering problem have been studied recently. In [6], the data gathering is assumed to be performed in *rounds* and each sensor can communicate

---

[*] Supported by the National Science Foundation under award No. IIS-0330445 and in part by DARPA under contract F33615-02-2-4005.

(in a single hop) with the base station and all other sensors. The total number of rounds is then maximized under a given energy constraint on the sensors. In [9], a non-linear programing formulation is proposed to explore the trade-off between energy consumed and the transmission bandwidth, which models the radio transmission energy according to Shannon's theorem. In [10], the data gathering problem is formulated as a linear programing problem and a $1 + \omega$ approximation algorithm is proposed. This algorithm further leads to a distributed heuristic.

Our study departs from the above with respect to the problem definition as well as the solution technique. For short-range communications, the difference in the energy consumption between sending and receiving data packets is almost negligible. We adopt the reasonable approximation that sending a data packet consumes the same amount energy as receiving a data packet [1]. The study in [9] and [10] differentiate the energy dissipated for sending and receiving data. Although the resulting problem formulations are indeed more accurate than ours, the improvement in accuracy is marginal for short-range communications.

In [6], each sensor generates exactly one data packet per round (a round corresponds to the occurrence of an event in the environment) to be transmitted to the base station. The system is assumed to be fully connected. The study in [6] also considers a very simple model of data aggregation where any sensor can aggregate all the received data packets into a single output data packet. In our sensor network model, each sensor communicates with a limited number of neighbors due to the short range of the communications, resulting in a general graph topology for the system. We study store-and-gather problems where data are locally stored on the sensors before the data gathering starts, and continuous sensing and gathering problems that models time critical applications. A unified flow optimization formulation is developed for the two classes of problems.

The constrained flow problem reduces to the standard network flow problem, which is a classical flow optimization problem. Many efficient algorithms have been developed ([2]) for the standard network flow problem. However, in terms of decentralization and adaptation (two properties desirable for WSNs), these well known algorithms are not suitable for data gathering problems in WSNs. In this paper, we develop a decentralized and adaptive algorithm for the maximum network flow problem. This algorithm is a modified version of the Push-Relabel algorithm [3]. In contrast to the Push-Relabel algorithm, it is adaptive to changes in the system. It finds the maximum flow in $O(n^2 \cdot |V|^2 \cdot |E|)$ time, where n is the number of adaptation operations, $|V|$ is the number of nodes, and $|E|$ is the number of links. With the algorithm thus established, we further develop a simple distributed protocol for data gathering. The performance of this protocol is studied through simulations.

The rest of the paper is organized as follows. The data gathering problems are discussed in Section 2. We show that these problems reduce to network flow problems with constraint on the vertices. In Section 3, we develop a mathematical formulation of the constrained network flow problem and show that it reduces to a standard network flow problem. In Section 4, we derive a relaxed form for the network flow problem. A distributed and adaptive algorithm is then developed for this relaxed problem. A simple protocol based on this algorithm is presented in Section 4.3. Experimental results are presented in Section 5. Section 6 concludes this paper.

2 Data Gathering Problems in Sensor Networks with Energy Constraint

2.1 Sensor Network Model

Suppose a network of sensors is deployed over a region. The location of the sensors are fixed and known *a priori*. The sensor network is represented by a graph $G(V, E)$, where V is the set of sensor nodes. $(u, v) \in E$ if $u \in V, v \in V$ and u is within the communication range of v. σ_u, the set of successors of u, is defined as $\sigma_u = \{v \in V | (u, v) \in E\}$. Similarly, ψ_u, the set of predecessors of u is defined as $\psi_u = \{v \in V | (v, u) \in E\}$. The event is sensed by a subset of sensors $V_c \subset V$. r is the base station to which the sensed data are transmitted. Sensors $S - V_c - \{r\}$ in the network does not sense the event but can relay the data sensed by V_c.

Data transfers are assumed to be performed via multi-hop communications where each hop is a short-range communication. This is due to the well known fact that long-distance wireless communication is expensive in terms of both implementation complexity and energy dissipation, especially when using the low-lying antennae and near-ground channels typically found in sensor networks [1]. Additionally, short-range communication enables efficient spatial frequency re-use in sensor networks [1].

Among the three categories (sensing, communication, and data processing) of power consumption, typically, a sensor node spends most of its energy in data communication. This includes both data transmission and reception. Our energy model for the sensors is based on the first order radio model described in [4]. The energy consumed by sensor u to transmit a k−bit data packet to sensor v is $T_{uv} = \varepsilon_{elec} \times k + \varepsilon_{amp} \times d_{uv}^2 \times k$, where ε_{elec} is the energy required for transceiver circuitry to process one bit of data, ε_{amp} is the energy required per bit of data for transmitter amplifier, and d_{uv} is the distance between u and v. Transmitter amplifier is not needed by u to receive data and the energy consumed by u to receive a k−bit data packet is $R_u = \varepsilon_{elec} \times k$. Typically, $\varepsilon_{elec} = 50nJ/bit$ and $\varepsilon_{amp} = 0.1nJ/bit/m^2$. This effectively translates to $\varepsilon_{amp} \times d_{uv}^2 \ll \varepsilon_{elec}$, especially when short transmission ranges ($\simeq 1m$) are considered. For the discussion in the rest of this paper, we adopt the approximation that $T_{uv} = R_u$ for $\forall\, u, v \in V$. We further assume that no data aggregation is performed during the transmission of the data.

We consider two classes of data gathering problems: store-and-gather problems and continuous sensing and gathering problems. An energy(power) budget B_u is imposed on each sensor node u. We assume that there is no energy constraint on base station r. We can normalize T_{uv} and R_u to 1. For the store-and-gather problems, B_u represents the total number of data packets that u can send and receive. For the continuous sensing and gathering problems, B_u represents the total number of data packets that u can send and receive in one unit of time. To simplify our discussions, we ignore the energy consumption of the sensors when sensing the environment. However, the rate at which sensor $u \in V_c$ can collect data from the environment is limited by the maximum sensing frequency g_u.

Communication link (u, v) has transmission bandwidth c_{uv}. We do not require the communication links to be identical. Two communication links may have different transmission latencies and/or bandwidth. Symmetry is not required either. It may be the case that $c_{uv} \neq c_{uv}$. If $(u, v) \notin E$, then we define $c_{uv} = 0$.

2.2 Store-and-Gather Problems

In store-and-gather problems, the information from the environment is sensed (possibly during a prolonged period) and stored locally at the sensors. These locally stored data are then transferred to the base station during the data gathering stage. This represents those data-oriented applications (e.g. counting the occurrences of endangered birds in a particular region) where the environment changes slowly. There is typically no deadline (or the deadline is loose enough to be ignored) on the duration of data gathering for such problems, and we are not interested in the speed at which the data is gathered. But due to the energy constraint, not all the stored data can be gathered by the base station, and we want to maximize the amount of data gathered.

For each $u \in V_c$, we assume that u has stored d_u data packet before the data gathering starts. Let $f(u, v)$ represent the number of data packets sent from u to v.

For the simplified scenario where V_c contains a single node s, we have the following problem formulation:

Single Source Maximum Data Volume (SMaxDV) Problem:
Given: A graph $G(V, E)$. Source $s \in V$ and sink $r \in V$. Each node $u \in V - \{r\}$ has energy budget B_u.
Find: A real valued function $f : E \to R$
Maximize: $\sum_{v \in \sigma_s} f(s, v)$
Subject to:

$$f(u, v) \geq 0 \qquad\qquad\qquad\qquad\qquad \text{for } \forall \, (u, v) \in E \qquad (1)$$
$$\sum_{v \in \sigma_u} f(u, v) + \sum_{v \in \psi_u} f(v, u) \leq B_u \qquad \text{for } u \in V - \{r\} \qquad (2)$$
$$\sum_{v \in \sigma_u} f(u, v) = \sum_{v \in \psi_u} f(v, u) \qquad \text{for } u \in V - \{s, r\} \qquad (3)$$

B_u is the energy budget of u. Since we have normalized both T_{uv} and R_u to 1, the total number of data packets that can be sent and received by u is bounded from above by B_u. Condition 2 above represents the energy constraint of the sensors. Sensors $V - \{s, r\}$ do not generate sensed data, nor should they posses any data packets upon the completion of the data gathering. This is reflected in Condition 3 above. We do not model d_s, the number of data packets stored at s before the data gathering starts. This is because d_s is an obvious upper bound for the SMaxDV problem, and can be handled trivially.

$|V_c| > 1$ represents the general scenario where the event is sensed by multiple sensors. This multi-source data gathering problem is formulated as follows:

Multiple Source Maximum Data Volume (MMaxDV) Problem:
Given: A graph $G(V, E)$. The set of source nodes $V_c \subset V$ and sink $r \in V$. Each node $u \in V - \{r\}$ has energy budget B_u. Each node $v \in V_c$ has d_v data packets that are locally stored before the data gathering starts.
Find: A real valued function $f : E \to R$
Maximize: $\sum_{v \in \sigma_s} f(s, v)$
Subject to:

$$f(u, v) \geq 0 \qquad\qquad\qquad\qquad\qquad\qquad \text{for } \forall \, (u, v) \in E \qquad (1)$$
$$\sum_{v \in \sigma_u} f(u, v) + \sum_{v \in \psi_u} f(v, u) \leq B_u \qquad\quad \text{for } u \in V - \{r\} \qquad (2)$$
$$\sum_{v \in \sigma_u} f(u, v) = \sum_{v \in \psi_u} f(v, u) \qquad\quad \text{for } u \in V - V_c - \{r\} \qquad (3)$$
$$\sum_{v \in \sigma_u} f(u, v) \leq \sum_{v \in \psi_u} f(v, u) + d_u \qquad \text{for } u \in S_c \qquad (4)$$

Similar to the SMaxDV problem, the net flow out of the intermediate nodes ($V - V_c - \{r\}$) is 0 in the MMaxDV problem, as is specified in Condition 3. For each source node $u \in V_c$, the net flow out of u cannot exceed the number of data packets previously stored at u. This is specified in Condition 4.

2.3 Continuous Sensing and Gathering Problems

The continuous sensing and gathering problems model those time critical applications that need to gather as much information as possible from the environment while the nodes are sensing. Examples of such applications include battle field surveillance, target tracking, etc. We want to maximize the total number of data packets that can be gathered by the base station r in one unit of time. We assume that the communications are scheduled by time/frequency division multiplexing or channel assignment techniques. We consider the scenario in which B_u is the maximum power consumption rate allowed by u. Let $f(u, v)$ denote the number of data packets sent from u to v in one unit of time.

Similar to the store-and-gather problem, we have the following mathematical formulation when V_c contains a single node s.

Single Source Maximum Data Throughput (SMaxDT) Problem:
Given: A graph $G(V, E)$. Source $s \in V$ and sink $r \in V$. Each node $u \in V - \{r\}$ has energy budget B_u. Each edge $(u, v) \in E$ has capacity c_{uv}.
Find: A real valued function $f : E \to R$
Maximize: $\sum_{v \in \sigma_s} f(s, v)$
Subject to:

$$0 \le f(u, v) \le c_{uv} \qquad \text{for } \forall (u, v) \in E \qquad (1)$$
$$\sum_{v \in \sigma_u} f(u, v) + \sum_{v \in \psi_u} f(v, u) \le B_u \qquad \text{for } u \in V - \{r\} \qquad (2)$$
$$\sum_{v \in \sigma_u} f(u, v) = \sum_{v \in \psi_u} f(v, u) \qquad \text{for } u \in V - \{s, r\} \qquad (3)$$

The major difference between the SMaxDV and the SMaxDT problem is the consideration of link capacities. In the SMaxDV problem, since there is no deadline for the data gathering, the primary factor that affects the maximum number of gathered data is the energy budgets of the sensors. But for the SMaxDT problem, the number of data packets that can be transferred over a link in one unit of time is not only affected by the energy budget, but also bounded from above by the capacity of that link, as is specified in Condition 1 above. For the SMaxDT problem, we do not model the impact of g_u because g_u is an obvious upper bound of the throughput and can be handled trivially.

Similarly, we can formulate the multiple source maximum data throughput problem as follows.

Multiple Source Maximum Data Throughput (MMaxDT) Problem:
Given: A graph $G(V, E)$. The set of source nodes $V_c \subset V$ and sink $r \in V$. Each node $u \in V - \{r\}$ has energy budget B_u. Each edge $(u, v) \in E$ has capacity c_{uv}.
Find: A real valued function $f : E \to R$
Maximize: $\sum_{v \in \sigma_s} f(s, v)$
Subject to:

$$0 \le f(u, v) \le c_{uv} \qquad \text{for } \forall (u, v) \in E \qquad (1)$$
$$\sum_{v \in \sigma_u} f(u, v) + \sum_{v \in \psi_u} f(v, u) \le B_u \qquad \text{for } u \in V - \{r\} \qquad (2)$$

$$\sum_{v\in\sigma_u} f(u,v) = \sum_{v\in\psi_u} f(v,u) \qquad\qquad \text{for } u \in V - V_c - \{r\} \quad (3)$$
$$\sum_{v\in\sigma_u} f(u,v) \le \sum_{v\in\psi_u} f(v,u) + g_u \qquad\qquad \text{for } u \in V_c \quad (4)$$

Condition 4 in the above problem formulation reflects the maximum sensing frequency of the sensors. By associating a certain amount of energy dissipation with each data packet sensed from the environment, we can even model the energy consumption in sensing the environment. Actually, our algorithm (discussed next) is capable of handling this scenario with a few minor modifications. The detailed discussion is omitted here due to space limitations.

3   Flow Maximization with Constraint on Vertices

3.1   Problem Reductions

In this section, we present the formulation of the constrained flow maximization problem where the vertices have limited capacities (CFM problem). The CFM problem is an abstraction of the four problems discussed in Section 2.

In the CFM problem, we are given a directed graph $G(V, E)$ with vertex set V and edge set E. Vertex u has capacity constraint $w_u > 0$. Edge (u, v) starts from vertex u, ends at vertex v, and has capacity constraint $c_{uv} > 0$. If $(u, v) \notin E$, we define $c_{uv} = 0$. We distinguish two vertices in G, source s, and sink r. A flow in G is a real valued function $f : E \to R$ that satisfies the following constraints:

1. $0 \le f(u,v) \le c_{uv}$ for $\forall\, (u, v) \in E$. This is the capacity constraint on edge (u, v).
2. $\sum_{v\in\sigma_u} f(u,v) = \sum_{v\in\psi_u} f(v,u)$ for $\forall\, u \in V - \{s, r\}$. This represents the flow conservation. The net amount of flow that goes through any of the vertices, except s and t, is zero.
3. $\sum_{v\in\sigma_u} f(u,v) + \sum_{v\in\psi_u} f(v,u) \le w_u$ for $\forall\, u \in V$. This is the capacity constraint of vertex u. The total amount of flow going through u cannot exceed w_u. This condition differentiates the CFM problem from the standard network flow problem.

The *value* of a flow f, denoted as $|f|$, is defined as $|f| = \sum_{v\in\sigma_s} f(s,v)$, which is the net flow that leaves s. In the CFM problem, we are given a graph with vertex and edge constraint, a source s, and a sink r, and we wish to find a flow with the maximum value.

It is straight forward to show that the SMaxDV and the SMaxDT problems reduce to the CFM problem. By adding a hypothetical super source node, the MMaxDV and the MMaxDT problems can also be reduced to SMaxDV and SMaxDT, respectively.

The CFM problem has been studied before [7]. Our focus in this paper is to design a distributed and adaptive algorithm for the proposed problem. It can be shown that the CFM problem reduces to a standard network flow problem. Due to the existence of condition 1, condition 3 is equivalent to $\sum_{v\in\sigma_u} f(u,v) \le w_u/2$ for $u \in V - \{s, r\}$. This means that the total amount of flow out of vertex u cannot exceed $w_u/2$. Suppose we split u ($u \in V - \{s, r\}$) into two nodes u_1 and u_2, re-direct all incoming links to u to arrive at u_1 and all the outgoing links from u to leave from u_2, and add a link from u_1 to u_2 with capacity $w_u/2$, then the vertex constraint w_u is fully represented by the capacity of link (u_1, u_2). Actually, such a split transforms all the vertex constraints to the corresponding link capacities, and effectively reduces the CFM problem to a standard network flow problem. A similar reduction can be found in [7].

The standard network flow problem is stated below:

Given: graph $G(V, E)$, source node $s \in V$, and sink node $r \in V$. Link (u, v) has capacity c_{uv}.

Maximize: $\sum_{v \in \sigma_s} f(s, v)$

Subject to:

$$0 \leq f(u, v) \leq c_{uv} \qquad \qquad \text{for } \forall\, (u, v) \in E \qquad (1)$$
$$\sum_{v \in \sigma_u} f(u, v) = \sum_{v \in \psi_u} f(v, u) \qquad \text{for } u \in V - \{s, r\} \qquad (2)$$

3.2 Relationship to Sensor Network Scenarios

The vertex capacity w_u in the CFM problem models the energy(power) budget B_u of the sensor nodes. B_u does not have to be the total remaining energy of u. For example, when the remaining battery power of a sensor is lower than a particular level, the sensor may limit its contribution to the data gathering operation by setting a small value for B_u (so that this sensor still has enough energy for future operations). For another example, if a sensor is deployed in a critical location so that it is utilized as a gateway to relay data packets to a group of sensors, then it may limit its energy budget for a particular data gathering operation, thereby conserving energy for future operations. These considerations can be captured by vertex capacity w_u in the CFM problem.

The edge capacity in the CFM problem models the communication rate (meaningful for continuous sensing and gathering problems) between adjacent sensor nodes. The edge capacity captures the available communication bandwidth between two nodes, which may be less than the the maximum available rate. For example, a node may reduce its radio transmission power to save energy, resulting in a less than maximum communication rate. This capacity can also vary over time based on environmental conditions. Our decentralized protocol results in an on-line algorithm for this scenario.

Because energy efficiency is a key consideration in WSNs, various techniques have been proposed to explore the trade-offs between processing/communication speed and energy consumption. This results in the continuous variation of the performance of the nodes. For example, the processing capabilities may change as a result of dynamic voltage scaling [8]. The data communication rate may change as a result of modulation scaling [11]. As proposed by various studies on energy efficiency, it is necessary for sensors to maintain a power management scheme, which continuously monitors and adjusts the energy consumption and hence changes the computation and communication performance of the sensors. In data gathering problems, these energy related adjustments translate to changes of parameters (node/link capacities) in the problem formulations. Determining the exact reasons and mechanisms behind such changes is beyond the scope of this paper. Instead, we focus on the development of data gathering algorithms that can adapt to such changes.

4   Distributed and Adaptive Algorithm To Maximize Flow

In this section, we first show that the maximum flow remains the same even if we relax
the flow conservation constraint. Then we develop a distributed and adaptive algorithm
based on this relaxation.

4.1   Relaxed Flow Maximization Problem

Consider the simple example in Figure 1 where s is the source, r is the sink, and u is the
only intermediate node. Obviously, the flow is maximized when $f(s, u) = f(u, r) = 10$.
Suppose s, u, and r form an actual system and s has sent 10 data packets to u. Then u
can send no more than 10 data packets to r even if we permit u to transfer more to r,
because u has received only 10 data packets. This means the actual system still works
as if $f(u, r) = 10$ even if we set $f(u, r) \geq 10$.

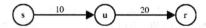

Fig. 1. An example of the relaxed network flow problem where $c_{su} = 10$ and $c_{ur} = 20$.

This leads to the following relaxed network flow problem:

Given: graph $G(V, E)$, source node $s \in V$, and sink node $r \in V$. Link (u, v) has
capacity c_{uv}.

Maximize: $|f| \stackrel{\text{def}}{=} \sum_{v \in \sigma_s} f(s, v)$

Subject to:

$$0 \leq f(u, v) \leq c_{uv} \qquad\qquad \text{for } \forall\, (u, v) \in E \qquad (1)$$
$$\sum_{v \in \sigma_u} f(u, v) \geq \sum_{v \in \psi_u} f(v, u) \qquad\qquad \text{for } u \in V - \{s, r\} \qquad (2)$$

Condition 2 differentiates the relaxed and the standard network flow problem. In the
relaxed problem, the total flow out of a node is larger than or equal to the total flow
into the node, and flow conservation is not required. The following theorem shows the
relation between the relaxed and the standard network flow problem.

Theorem 1. *Given graph $G(V, E)$, source s and sink r. If f^* is an optimal solution to the
relaxed network flow problem, then there exists an optimal solution f' to the standard
network flow problem such that $f'(u, v) \leq f^*(u, v)$ for $\forall\, (u, v) \in E$. Additionally,
$|f^*| = |f'|$.*

Proof of the theorem is not difficult and omitted here due to space limitations. If we
interpret $f^*(u, v)$ as the number of data units that we ask u to transfer and $f'(u, v)$ as the
number of data units that u actually transfers, then this theorem essentially indicates that
the solution to a relaxed flow problem can have an actual implementation that satisfies
flow conservation!

4.2   The Algorithm

In this section, we develop a decentralized and adaptive algorithm for the relaxed network flow maximization problem. This algorithm is a modified version of the Push-Relabel algorithm [2] and is denoted as the *Relaxed Incremental Push-Relabel* (RIPR) algorithm.

The Push-Relabel algorithm is a well known algorithm for network flow maximization. It has a decentralized implementation where every node only needs to exchange messages with its immediate neighbors and makes decisions locally. But in order to be adaptive to the changes in the system, this algorithm has to be re-initialized and re-run from scratch each time when some parameters (weight of the nodes and edges in the graph) of the flow maximization problem change. Each time before starting to search for the new optimal solution, the algorithm needs to make sure that every node has finished its local initialization, which requires a global synchronization and compromises the property of decentralization.

In contrast to the Push-Relabel algorithm, our algorithm introduces the adaptation operation, which is performed upon the current values of $f(u,v)$ and $h(u)$ for $\forall u, v \in V$. In other words, our algorithm performs *incremental* optimization as the parameters of the system change. Our algorithm does not need global synchronizations. Another difference is that our algorithm applies to the relaxed network flow problem, rather than the standard one.

An integer valued auxiliary function $h(u)$ is defined for $u \in V$, which will be explained in the algorithm. The algorithm is as follows:

1. *Initialization:* $h(u)$, and $f(u,v)$ are initialized as follows:

$h(u)$	$= 0$	for $\forall\, u \in V$
$f(u,v)$	$= 0$	for $\forall\, u, v \in E$
$h(s)$	$= |V|$	
$f(s,v)$	$= l_{sv}$	for $\forall\, v \in V$
$f(v,s)$	$= -l_{vs}$	for $\forall\, v \in V$
$e(u)$	$= \sum_{v \in V} f(v,u)$	for $\forall\, u \in V$

2. *Search for maximum flow:*
   Each node $u \in V - \{s, r\}$ conducts one of the following three operations as long as $e(u) \neq 0$:

   a) *Push(u, v):* applies when $e(u) > 0$ and $\exists (u, v) \in E_f$ s.t. $h(u) > h(v)$,

d	$= \min(e(u), c_f(u,v))$
$f(u,v)$	$= f(u,v) + d$
$f(v,u)$	$= -f(u,v)$
$e(u)$	$= e(u) - d$
$e(v)$	$= e(v) + d$

   b) *Relabel(u):* applies when $e(u) > 0$ and $h(u) \leq h(v)$ for $\forall\, (u, v) \in E_f$,

   $$h(u) = \min_{(u,v) \in E_f} h(v) + 1$$

3. *Adaptation to changes in the system:* For the flow maximization problem, the only possible change that can occur in the system is the increase or decrease of the capacity of some edges. Suppose the value of l_{uv} changes to l'_{uv}, the following four scenarios are considered when performing the Adaptation (u, v) operation:

a) if $l'_{uv} > l_{uv}$ and $f(u,v) < l_{uv}$, do nothing.

b) if $l'_{uv} > l_{uv}$ and $f(u,v) = l_{uv}$, then

$$
\begin{aligned}
h(s) &= h(s) + 3|V| \\
f(s,v) &= l_{sv} &&\text{for } \forall\, v \in \sigma_s \\
f(v,s) &= -l_{sv} &&\text{for } \forall\, v \in \sigma_s \\
e(v) &= \textstyle\sum_{v \in V} f(v,v) &&\text{for } \forall\, v \in V
\end{aligned}
$$

c) if $l'_{uv} < l_{uv}$ and $f(u,v) \le l_{uv}$, do nothing.

d) if $l'_{uv} < l_{uv}$ and $f(u,v) > l_{uv}$, then

$$
\begin{aligned}
h(s) &= h(s) + 3|V| \\
f(s,v) &= l_{sv} &&\text{for } \forall\, v \in \sigma_s \\
f(v,s) &= -l_{sv} &&\text{for } \forall\, v \in \sigma_s \\
e(v) &= \textstyle\sum_{v \in V} f(v,v) &&\text{for } \forall\, v \in V \\
f(u,v) &= l'_{uv} \\
f(v,u) &= -l'_{uv} \\
e(u) &= e(u) + (l_{uv} - l'_{uv}) \\
e(v) &= e(v) - (l_{uv} - l'_{uv})
\end{aligned}
$$

In the above algorithm, the 'adaptation' is activated when some link capacity changes in the relaxed flow problem. Because link capacities in the relaxed flow problem map to either vertex or link capacities in the corresponding CFM problem, the adaptation operation actually reacts to capacity changes in both vertex and link capacities. The 'Push' and 'Relabel' operations are called the *basic operations*. Every node in the graph determines its own behavior based on the knowledge about itself and its neighbors (as can be seen, the Push and Relabel operations are triggered by variables whose value are locally known by the nodes). No central coordinator or global information about the system is needed. More importantly, unlike the Push-Relabel algorithm, no global synchronization is needed when the RIPR algorithm adapts to the changes in the system.

An intuitive explanation of the RIPR is as follows. $h(u)$ represents, intuitively, the shortest distance from u to t when $h(u) \le h(s)$. When $h(u) > h(s)$, $h(u) - h(s)$ represents the shortest distance from u to s. Hence the RIPR algorithm attempts to push more flow from s to t along the shortest path; excessive flow of intermediate nodes are pushed back to s along the shortest path. Similar to the Edmonds-Karp algorithm[2], such a choice of paths can lead to an optimal solution. This is formally stated in the next theorem.

Theorem 2. *The RIPR algorithm finds the maximum flow for the relaxed flow problem with $O(n^2 \cdot |V|^2 \cdot |E|)$ basic operations, where n is the number of adaptation operations performed, $|V|$ is the number of nodes in the graph, and $|E|$ is the number of edges in the graph.*

The proof of Theorem 2 consists of two parts. First, we prove that if the algorithm terminates, it obtains the maximum flow. Next, we prove that the algorithm does terminate by finding upper bounds on the number of Push and Relabel operations that are performed. Details of the proof are omitted here due to space limitations and can be found in [5].

At the time of the preparation of this paper, we can prove that the RIPR algorithm needs $O(n^2 \cdot |V|^2 \cdot |E|)$ basic operations to find the maximum flow. However, experiments

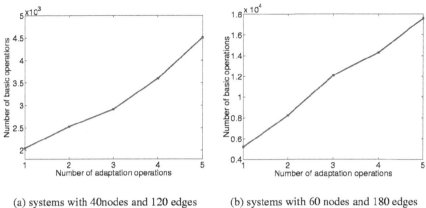

(a) systems with 40nodes and 120 edges   (b) systems with 60 nodes and 180 edges

Fig. 2. Complexity of the RIPR algorithm. Number of basic operations vs number of adaptations

show that the average performance is better than the bound stated in Theorem 2. Figure 2 shows the experimental results. Both Figure 2 (a) and (b) report results averaged over 50 randomly generated systems. The x-axis represents the number of adaptation operations. The y-axis shows the average number of basic operations performed by the algorithm to find the maximum flow. These simulation results suggest that, instead of n^2, there may be a linear relation between the total number of basic operations and the number of adaptations. We are currently investigating the possibility of a better complexity bound.

4.3   A Simple Protocol for Data Gathering

In this section, we show a simple on-line protocol for SMaxDT problem based on the RIPR algorithm in Section 4.

The maximum flow obtained using the RIPR algorithm does not tell us how to transfer the data through the network. It only contains information such as 'u needs to transfer 0.38 unit of data to v in one unit of time'. But the actual system needs to deal with integer number of data packets. Furthermore, before the RIPR algorithm finds the maximum flow, a node, say u, may have a positive valued $e(u)$, which means u is accumulating data at that time instance. Yet such a node u should not keep accumulating data as $e(u)$ will eventually be driven to zero.

These issues are addressed by maintaining a data buffer at each node. Initially, all the data buffers are empty. The source node s senses the environment and fills its buffer continuously. At any time instance, let β_u denote the amount of buffer used by node u. Each node $u \in V$ operates as follows:

1. Contact the adjacent node(s) and execute the RIPR algorithm.
2. While $\beta_u > 0$, send message 'request to send' to all successors v of u s.t. $f(u, v) > 0$. If 'clear to send' is received from v, then set $\beta_u \leftarrow \beta_u - 1$ and send a data packet to v.

3. Upon receiving 'request to send', u acknowledges 'clear to send' if $\beta_u \leq U$. Here U is a pre-set threshold that limits the maximum number of data packets a buffer can hold.

For node s, it stops sensing if $\beta_s > U$. Two types of data are transferred in the system: the control messages that are used by the RIPR algorithm, and the sensed data themselves. The control messages are exchanged among the nodes to query and update the values of $f(u,v)$ and h_u etc. The 'request to send' and 'clear to send' messages are also control messages, though they are more related to data transfer. The control messages and the sensed data are transmitted over the same links and higher priority is given to the control messages in case of a conflict.

For the MMaxDT problem, the situation is a bit more complicated. Since the MMaxDT problem is reduced to the SMaxDT problem by adding a hypothetical super source node s', the RIPR algorithm needs to maintain the flow out of s' as well as the value of function $h(s')$. Additionally, the values of $f(s',v)$ ($v \in V_c$) and $h(s')$ are needed by all nodes $u \in V_c$ during the execution of the algorithm. Because s' is not an actual sensor, sensors in V_c therefore need to maintain a consistent image of s'. This requires some regional coordination among sensors in V_c and may require some extra cost to actually implement such a consistent image.

SMaxDV and MMaxDV are by nature off-line problems and hence we do not develop on-line protocols for them. Actually, we can approximate the optimal solution to SMaxDV and MMaxDV using some simple heuristics. For example, we can find the shortest path (in terms of the number of hops) from the source (or one of the sources) to the sink and push as many data packets as possible along the path until the energy of some node(s) along the path is depleted. Then we find the second shortest path and push data packets along this path. This is repeated till the source and sink are disconnected. Such a heuristic is similar to the augmenting path method for the standard flow problem except that the paths are searched in the original graph, rather than the residual graph. The heuristic may not find the optimal solution. However, it can be shown experimentally that such a heuristic can achieve a close to optimal performance [5].

5   Experimental Results

Simulations were conducted to illustrate the effectiveness of the proposed data gathering algorithm. We focus on the continuous sensing and gathering problems where our simple protocol can be applied.

As we have already proved that the RIPR algorithm finds the maximum flow, we do not show here the throughput achievable by our protocol, which is based on the RIPR algorithm. Actually, throughput up to 95% of the optimal has been observed in the simulations.

As claimed in Section 4, adaptation is an important properties of the proposed algorithm and hence the data gathering protocol. This is illustrated in the following simulations of the SMaxDT problem.

The simulated sensor network was generated by randomly scattering 40 sensor nodes in a unit square. The base station was located at the left-corner of the square. The source node was randomly chosen from the sensor nodes. The number of sensor nodes that

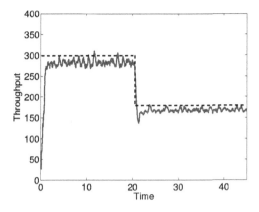

Fig. 3. Adaptation to changes in the system.

can communicate directly to a specific node is determined by a connectivity parameter, $\kappa \in (0, 1]$, such that the average number of neighbors of a sensor node is $200\pi\kappa^2$. B_u's are uniformly distributed between 0 and B_{max}. B_{max} was set to 500 for the simulations. We assume a signal decaying factor of r^{-2}. the flow capacity between sensor nodes u and v is determined by Shannon's theorem as $l_{uv} = W \log(1 + \frac{P_{uv}r_{uv}^{-2}}{\eta})$ where W is the bandwidth of the link, r_{uv} is the distance between u and v, P_{uv} is the transmission power on link (u, v), and η is the noise in the communication channel. In all the simulations, W was set to $1KHz$, P_{uv} was set to $10^{-3}mW$, and η was set to $10^{-6}mW$. U was set to 2.

During the course of data gathering, the capacities of a set of randomly selected edges were reduced by 50% at time instance $t = 20s$. The energy budgets of another set of randomly selected sensors were reduced by 30% at $t = 20s$. The achieved data gathering throughput is shown in Figure 3 , where the dotted line represents the optimal throughput calculated offline. Let $N(t)$ represents the total number of data packets received by the base station from time 0 to time instance t. The instantaneous throughput (reported in Figure 3) at time instance τ is calculated as $(N(\tau + 0.1) - N(\tau - 0.1))/0.2$. Various randomly generated systems were simulated under such settings. The same adaptation trend was observed and only one such result is shown in Figure 3. When the system parameters changed at $t = 20s$, the adaptation procedure was activated and the data gathering was adapted. As can be seen, the system operates at (close to) the new optimal throughput after the adaptation was completed.

6   Conclusion

In this paper, we studied a set of data gathering problems in energy-constrained wireless sensor networks. We reduced such problems to a network flow maximization problem with vertex capacity constraint, which can be further reduced to the standard network flow problem. After deriving a relaxed formulation for the standard network problem, we developed a distributed and adaptive algorithm to maximize the flow. This algorithm can be implemented as a simple data gathering protocol for WSNs.

One of the future directions is to design distributed algorithms that do not generate excessive flow at the nodes (i.e. $e(u)$ does not become positive) during the execution. Our formulation of constrained flow optimizations can be applied to problems beyond those four discussed in this paper. For example, the system model considered in [6] gathers data in rounds. In each round, every sensor generates one data packet and the data packets from all the sensors need to be collected by the sink. The goal is to maximize the total number of rounds the system can operate under energy constraints of the nodes. This problem can be described by our constrained flow formulation and an optimal solution can be developed [5].

Acknowledgment. We thank the anonymous reviewer for pointing out earlier work on constrained flow problems.

References

1. I. F. Akyildiz, W. Su, Y. Sankarasubramaniam, and E. Cyirci. Wireless Sensor Networks: A Survey. *Computer Networks*, 38(4):393–422, 2002.
2. T. H. Cormen, C. E. Leiserson, and R. L. Rivest. *Introduction to Algorithms*. MIT Press, 1992.
3. A. V. Goldberg and R. E. Tarjan. A New Approach to the Maximum Flow Problem. *Journal of Association for Computing Machinery*, 35:921–940, 1988.
4. W. R. Heinzelman, A. Chandrakasan, and H. Balakrishnan. Energy Efficient Communication Protocol for Wireless Micro-sensor Networks. In *Proceedings of IEEE Hawaii International Conference on System Sciences*, 2000.
5. B. Hong and V. K. Prasanna. Constrained Flow Optimization with Applications to Data Gathering in Sensor Networks. Technical Report CENG-2004-07, Department of Electrical Engineering, University of Southern California, http://www-scf.usc.edu/~bohong/report_may04.ps, May 2004.
6. K. Kalpakis, K. Dasgupta, and P. Namjoshi. Maximum Lifetime Data Gathering and Aggregation in Wireless Sensor Networks. *IEEE Networks '02 Conference*, 2002.
7. E. L. Lawler. *Combinatorial Optimization : Networks and Matroids*. Holt, Rinehart and Winston, 1976.
8. R. Min, T. Furrer, and A. Chandrakasan. Dynamic Voltage Scaling Techniques for Distributed Microsensor Networks. *Workshop on VLSI (WVLSI '00)*, April 2000.
9. F. Ordonez and B. Krishnamachari. Optimal Information Extraction in Energy-Limited Wireless Sensor Networks. *to appear in IEEE Journal on Selected Areas in Communications, special issue on Fundamental Performance Limits of Wireless Sensor Networks, 2004.*
10. N. Sadagopan and B. Krishnamachari. Maximizing Data Extraction in Energy-Limited Sensor Networks. *IEEE Infocom 2004*, 2004.
11. C. Schurgers, O. Aberthorne, and M. Srivastava. Modulation Scaling for Energy Aware Communication Systems. In *International Symposium on Low Power Electronics and Design*, pages 96–99, 2001.

Author Index

Lecture Notes in Computer Science

For information about Vols. 1–3031

please contact your bookseller or Springer-Verlag